The Seasons of
Fire

ENVIRONMENTAL ARTS AND
HUMANITIES SERIES

The Seasons of
Fire

REFLECTIONS ON
Fire in the West

David J. Strohmaier

UNIVERSITY OF NEVADA PRESS

Reno & Las Vegas

Environmental Arts and Humanities Series
Series Editor: Scott Slovic

University of Nevada Press,
Reno, Nevada 89557 USA
Copyright © 2001 by David J. Strohmaier
All rights reserved.
Manufactured in the United States of America
Design by Kaelin Chappell

Library of Congress Cataloging-in-Publication Data
Strohmaier, David J. (David Jon), 1965–
The seasons of fire : reflections on fire in the West /
David J. Strohmaier.
p. cm.—(Environmental arts and humanities series)
Includes bibliographical references and index.
ISBN 0-87417-483-x (pbk. : alk. paper)
1. Wildfires—Prevention and control—Northwest, Pacific.
2. Fire ecology—Northwest, Pacific.
3. Strohmaier, David J. (David Jon), 1965–
4. Wildfire fighters—Northwest, Pacific.
I. Title. II. Series.
SD421.32.N67 S77 2001
634.9'618'0978—dc21
2001001406

The paper used in this book meets the requirements of
American National Standard for Information Sciences—
Permanence of Paper for Printed Materials,
ANSI Z39.48-1984. Binding materials were selected
for strength and durability.

First Printing
10 09 08 07 06 05 04 03 02 01
5 4 3 2 1

For Mom and Dad,
and for Gretchen

For everything there is a season,
and a time for every matter under heaven.

—QOHELETH, Ecclesiastes 3:1

CONTENTS

ACKNOWLEDGMENTS

I owe many thanks to those friends, family, and colleagues who helped kindle, tend, and engage the ideas that became this book. My thanks to:

Richard Fern, Chris Hice, Jennifer Collins, and Mike Oyster, who provided valuable feedback on early versions of the manuscript; Robert Standley, Steve Wolfe, and Mike Oyster, whose companionship around our fire rings of spring and fall, along trout-laden waters and mahogany-studded mountain ridges, inspired many of my reflections on fire; my friends and co-workers with the Prineville District of the Bureau of Land Management, particularly the fire crews at the Bakeoven and Grass Valley guard stations, who stand behind many of the narratives in this book; Don Snow and the students in the University of Montana's 1998 environmental writing workshop; Bill Kittredge and the participants in the 1999 Environmental Writing Institute; and Christopher Preston, for insightful comments on selected chapters.

I thank my parents, Ed and Ella, who gave me encouragement to think and write and engage the natural world, and whose home fires gave and sustained life.

Most importantly, I thank Gretchen—my friend, companion, and wife—who breathed life into these pages when they began to cool, fueled many lively conversations with her keen insight, patiently read countless versions of the manuscript, and never abandoned hope that this project would eventually come to print.

PROLOGUE

Fire warms, dissolves, enlightens;
is the great promoter of vegetation and life,
if not necessary to the support of both.

—WILLIAM PALEY, "On the Elements"

Fire is a happening, a dynamic event, smeared across space and time and encompassing human as well as physico-chemical dimensions.[1] Aspects of this happening may be easy enough to quantify in terms of British thermal units, flame lengths, or degrees Fahrenheit and Celsius, but the broader significance that fire has for human life—not to mention its diverse ecological roles—is more elusive. Fire is not a mere thing that can be examined in isolation, as if you could pick it up, turn it over, palpate its contours, and then set it down again.

To exist, fire must remain in motion, which adds to its ethereal character. In a wildfire, flames stretch vertically into the atmosphere and fan horizontally over the landscape. Motion, however, is just as real for the penned-up fire in a cast-iron woodstove. Even there, fire sweeps across the surface of split wood, igniting vapors in a swirl of flames that dance their way around the firebox: a *happening*.

Besides movement through space and time, the dynamics of this happening also include life. Terrestrial fire requires fuel, and fuel comprises vegetation that once was, or still is, alive. To the extent that animal life is directly affected by what, and how, fires burn, it is also a part of this happening. Human beings, too, are intimately bound to fire. We live in both the season of fire under open sky and the season of fire ensconced as contained warmth, and for the most part, we maintain enough continuity in our lives to recognize ourselves as the same persons in each. As a result, in any solar year, we straddle several generations, several seasons of fire—fire that has a life and environment all its own yet, at times, has a life very dependent on us for its continued being and coming to be: a symbiotic life on which, maybe more than we realize, we too depend.

In a very real sense, our motions contour the equinoctial movements of fire and have done so for a very, very long time. Countless millennia ago, fire from the heavens connected to earthen tinder, and wildfire was born. In geologic time, it's only been recently that our ancestors began playing with embers.[2] And with modernity's ability to manipulate fire through technology and industry has come a corresponding desire to control and rein in native fire— motivated by fear, by revulsion toward that which destroys, and by indifference toward the prosaic roles that fire has filled in nature. Nevertheless, as long as humanity holds an unreflective antagonism toward fire, we will fight fire in the wildlands with a vengeance. It will remain anathema for anyone who engages fire on the pseudo-battlefields of forest and mesa to sound sympathetic toward the alleged enemy; this would be the height of treason. But, despite our efforts to the contrary, the fires of summer will come, as they always have and inevitably will, and sympathetic I am. Fire in the wildlands, taken as an event, does destroy life; seen as a process, it brings forth life. As both event and process, fire may even be a thing of beauty. This book is an apologetic on behalf of fire and on behalf of those who find value in watching, tending, and engaging fire.

The term "season" can be applied to multiple subjects. To speak of a season is to speak of an unfolding period of time, identifiable though dynamic: A season is in motion. Some seasons recur in endless cycles of renewal; other seasons unfold toward a terminus. The seasons of the year press forward from spring to summer to fall to winter and to spring once again, seemingly forever. A final winter will come to the earth billions of years hence, so we're told, but this is a cold abstraction. Human life is also marked by seasons. Our lives are analogous to the annually renewed flora that sprouts, matures, then goes dormant or dies. We are born, we grow, and we mature. Eventually we slow, weaken, and die. However, the winter—the terminus—of human life is no abstraction; it is a hard reality.

Like the seasons of our lives and of the year, fire also has seasons. It manifests itself in those discrete solar periods, identifiably different in each. The fires of spring are not the fires of summer, fall, or winter. The fire that is nature's offspring is different from that kindled by human hands; it has a seasonal beginning, middle, and end—the spring of fire, the summer of fire, the fall and winter of fire. But with the help of humans, fire can exist even during winter, beyond the grave.

Any person who has ever fought fire or stoked a glowing hearth realizes that human attraction to fire transcends our mortal (somewhat illusory) ability to control it. The wooden match, the pilot light, or an automobile's engine are all human-initiated means to produce combustion. Well and good. But there is another dimension to fire. The fire that blooms under open sky and unlocks the growth rings of nature's bounty can be more than useful; it has the potential to incite reverie, which for many embraces the spiritual. Fire has much to reveal if we will only listen to it and stare intently into its flame. We probably do so already and don't even realize it, or we realize it and feel ashamed or incapable of knowing how to articulate our feelings. So besides offering an apologetic for fire's significant place in nature throughout the seasons of the year, I offer a hearth upon which to gaze, revisiting how fire in large and small ways has played, and might still play, a role in our lives.

To grasp the seasonal nature of fire, of our lives, and of our relationship to fire is no idle exercise. Rather, to be *present to* fire, or in the *presence of* fire, in all its moods and modalities, has the potential to illumine who we are and what we are about in a sometimes morally ambiguous world. There is no *Feuer* animism, no religion of the flame. To stare deeply into the embers of a gasping fire is to behold in a microcosm a miracle latent in our terrestrial home: There is fire in the wood and the wood has become fire, and there is more under the sun than meets our eyes.

In the following pages, I engage in a project of remembering— remembering how fire, in one form or another, clings to the diurnal and seasonal unfolding of our years, and how this attachment can be deeply meaningful. The chapters span the seasons of a year, the fire within those seasons, and the experience of some whose lives have been shaped by fire. The subject of fire and humanity's interaction with it has many entry points: historical, sociological, ecological, psychological, philosophical, and religious. Mine is a project of generality, though generality lodged in the specificity of the seasons of fire I've experienced as a wildland firefighter and inhabitant of western North America. I write as one who has engaged the fires of summer in forest and range—in their infancy and on their deathbed—and done so primarily on the northern cusp of America's Great Basin. I also write as one who has watched and tended fires, during the wet of spring and the cold of winter, from the perimeter of a rock ring or a comfortable distance from a glowing hearth. This hearth that I offer the reader is contoured by rimrock and dyed with the hues of semiarid horizons, benchmarked by what I've come to see as the significant cairns marking the life of fire and the lives of humans who have been touched by fire. This is a book for anyone who has ever looked into the face of a campfire and reveled, or who has ever smelled the bittersweet smoke of juniper and pine and said with a deep sigh, "Ah . . . very nice."

◆   ◆   ◆

By incremental movements—sometimes smooth and imperceptible, sometimes jolting—first autumn, and now winter, has enveloped my neighborhood. Out my study window I see a man stoking yard debris in a rusted, fifty-five-gallon drum. So many sticks and branches fill the drum's jagged mouth that some limbs, at the top of the spring-loaded tangle, reach vertically two or three feet. Occasionally, after the fire has gnawed a path up the limbs, the fire-weakened branches bend over and snap off around their spindly midriffs, leaving a circle of half-burnt sticks heaped up around the base of the ribbed, reddish brown cylinder. By the look of the blackened ground around the barrel, this is a common occurrence. Bright-orange fingers poke through rusted-out holes in the barrel's side, while blue-gray smoke belches up, splitting lazily and rejoining as it arcs in the canopy of a leafless cottonwood. The smell of melted plastic and smoldering egg cartons lies thick in the air.

Sure enough, the barrel is a way to avoid paying someone to haul off branches, twigs, and trash. But maybe my neighbor has additional reasons for inciting this barrel-born conflagration, like *The landfill is getting too full* or *I'll add the ashes to the garden next spring.* Then again, maybe he just likes to see it burn, because this is the time of year you burn branches, the fire warms your fingers, and the smoke of poplar and cottonwood is kind of nice.

Spring

I

Buildup

The reversals of fire: first sea;
but of sea half is earth,
half lightning storm.

—HERACLITUS, Fragment 38

The knobby, green metal posts drive easily into the sandy soil.
I never have liked pounding fence posts and stringing wire, but
it keeps you on the payroll when fire is hibernating and unem-
ployment checks run dry. With practice, a person could get
good at fencing, I suppose. But I neither practice enough nor
have the requisite passion to convert these knuckle-bloodying
rolls of barbed wire into tight, geometric lines.

It's April, and a half-dozen other preseason firefighters and I
are at work building a small exclosure around a guzzler. The
fence, according to our district Bureau of Land Management
(BLM) wildlife biologist, is supposed to exclude cows yet allow
deer and antelope access to a glorified birdbath. The guzzler
consists of a corrugated metal rack that catches rainfall, fun-
neling it into a buried storage tank that judiciously doles it out

to a little concrete basin during the rest of the year. While it does rain in this desert steppe country—coming mainly in the form of spring and early summer cloudbursts—water is limited. Still, our creating these little oases in Central Oregon's Millican Valley seems sort of suspect. Lord knows how the critters out here wetted their parched lips for the eons before land managers and idle firefighters showed up to create fenced-in birdbaths. And today, after we've driven through chocolaty sludge and water-filled ruts to get here, fencing guzzlers seems less than urgent.

For some weeks, the weather lumbering across the high desert has been entrenched in indecisiveness. One day the temperature never ascends above freezing; the next, it's a monotonous fifty-four degrees, dipping to an equally lukewarm forty-five at night. For the most part, it's dry, if overcast. Yet sometimes rain comes generous and long, as Pacific frontal systems whip inland, clawing over the coast and Cascade ranges, still packing enough energy to soak the high desert. Our half-amphibious journey to today's guzzler affirms this.

Pine Mountain rises to the south. A little further to the east are Studhorse and Frederick Buttes, our BLM District boundary, and a clear horizon of sage, Idaho fescue, cheatgrass, occasional junipers, cow shit, and ribbons of tilted rimrock. The routine is familiar—two-hour drive to the work site, one hour of work, lunch, one hour of work, two-hour drive home. Except today it's different. At the edge of the sky they come: small chunks of cauliflower-like clouds peaking up from behind the yellow-and-gray landscape.

The weather report hinted at this: "Chance of afternoon thunderstorms." As the morning wears on, the forecast—amazingly—looks accurate. Somewhere to our southwest, over the southern end of the fir-draped Cascades, up over the Siskiyous, and then a hundred or so miles out over the Pacific, a giant pinwheel spins. Lifting and grinding toward the west, south, east, and northeast, it pivots like the hands of a giant clock running backward. Every spring, this meteorological curiosity forms and jets just enough moisture inland, combined with the sun's warmth, that clouds form and coalesce over the western edge of America's Great Basin. A quick glance at a televised weather report's satellite map tells more: There, off the Oregon-California coastline, sits a cloudy spiral design (often marked on these maps with an L in its center for "low pressure").

What the satellite imagery doesn't reveal, however, is that if enough heat and moisture and instability are present, this might be New Year's Day. Unlike most people who mark the beginning of the new year with either a giant ball dropping in Times Square or a holiday punctuated by flying pigskins, those who will don soot- and oil-darkened boots and take up pulaskis during June mark the new year with the first sighting of a thunderhead—cumulonimbus. (A pulaski is a combination grub hoe–ax.) This new year can be marked on a calendar, but only after the fact. Unlike the particularity of the first of January, which arrives right on schedule year after year, the year of the firefighter, and of fire, is changeable. Sometime in March, April, or May, the pinwheel materializes over the Pacific and begins to spin. A chain of events unfurls, eventuating in white puffs suspended in a sea of blue and, finally, in gray wisps of smoke rising from rusty punk, duff, and splintered bark.

Satellite photos and forecasts confirm what one already knows; they lend credence to what one suspects. Which is to say, the meteorologist isn't needed to herald the new year. There was a day, not too long ago, when looking down on our earthen home from the heavens was not possible. Then, one looked to the south.

I try to avoid watching the spontaneously generating clouds

too intently. Not only would I likely smash someone's hand with the post pounder, but, like many objects of anticipation, the clouds may never come. Rightly anticipating cumulonimbus's arrival is truly an exercise in inattentiveness. This is the same phenomenon known all too well to anglers entranced with awe-inspiring backlashes of monofilament, pearly strings of wind knots welded into 6X tippet, or yards of fly line lassoed around their waders. The minute you allow such a mess of slack line to drift unattended in the current while you study such anomalies, something beneath the current takes up the slack. In a similar paradox, no sooner do you take your eyes off the budding cauliflower patch than you find yourself surrounded by row upon row of puffy whiteness. It's still premature to gaze too intently at this atmospheric cornucopia; it must do much more growing before the old year in fact ends and another year begins.

By midafternoon, the garden-variety cauliflower have transformed into busts of creatures from the animal kingdom. Giant giraffes with crenellated necks erupt thousands of feet vertically. At some point, maybe that at which they detect a delectable fragrance of the new year, the milky necks bend over and angle off downwind, north by northeast. Such transmogrifications are multiplied all across the migrating herd of cumulus, above their sage pasturelands.

I recall sitting through a unit on weather in my high school geography class. Assisted by sheer force of will and rote memorization, I regurgitated on fill-in-the-blank quizzes the names of common cloud formations: stratus, cirrus, cirrostratus, cumulus, cumulonimbus (or as I probably thought at the time about my teacher, "come-u-ol'-numb-ass"). But these categories meant little until my literal baptism by fire. Once I was initiated into the family of those who extinguish fire, clouds became both icons—pointing to times, seasons, and realities beyond themselves—and things of beauty, to be appreciated in and of themselves.

The cumulo-herds of giraffes or horses or whatever long-necked animals one finds projected skyward grow and build, and the animals thrust their heads into the sky-blue firmament. Like elk during the rut, their necks swell; simultaneously, vertical development continues, and some thickening spires even coalesce and become two-headed beasts. The skins of these celestial creatures remain cauliflower-like, broken up by dark creases. The crowns of their heads are pure, luminescent white. These clouds are certainly different from those that flowed west to east during winter months. Those came in waves, pouring out across the desert like a film, even if occasionally vivified by a thunderhead, incognito. Today, these clouds erupt. They appear solid, framed by sharp contrasts. Not only are the brilliant white tops juxtaposed against pure blue, but each towering side is pressed in and up by hard glacial blue. I've seen storms in the Midwest, and they, too, grow like these. Yet those Midwestern storms have always struck me as having a certain muggy haziness about them. Maybe this is a trait of the particular storms that I saw or the times of the year I observed them; maybe the strong frontal storms of that region are more humid than their Western cousins. At any rate, here in the high desert, the pinwheel to the southwest spins within clear boundaries. And this all happens much quicker if you aren't watching.

Having completed our standard two hours of fencing for the day, we slog our way back to pavement, and on to our headquarters in Prineville. There, in our fire equipment cache, we wait next to the phone, just in case New Year's arrives and Dispatch has an incident for us to take action on, encrypted in the form of a legal description—township, range, section, and quarter quarter-section.

By now, the legion herds are gone. There are still a number of necks piled up and sniffing toward the north; however, amidst these few adolescents reside even larger creatures— beings of a new sort. All their whiteness has drained upward, leaving their bases shades of gray and black. What looks like bunches of grapes, or maybe upside-down skeins of salmon eggs, ripples the cloud bottoms: mammatus, as the meteorologists call them. The cloud is ripe for something to happen. By this time, the heads of these behemoths sheer off into great anvils, and the whole mass continues migrating north, over Millican, Horse Ridge, Rodman Rim, and the upper Crooked River.

At five o'clock in the afternoon, the darkness of midnight arrives. A spark flashes between the grapes and the desert floor, and the sound of a striking anvil echoes across mesas, against rimrock, and through arroyo. New Year's.

The year is only in its infancy. And so is its fire. Most likely, even if the spiderweb-like stringers of lightning connect with something dry enough to burn, the blaze will be short-lived as the grapes burst and send down curtains of rain. These storms are wet. At first, in this early spring drama, dark, wispy brush strokes trail off beneath cloud bases, indicating virga, precipitation that evaporates before it hits the ground. Soon the brush strokes transform into curtains of thick blackness. Nevertheless, whatever the degree of sogginess in these incipient spring storms, Cumulonimbus has arrived, and, with it, the season of fire in the wildlands. This is a definable moment. Like fire, it is a happening: an unfolding of latent energies that are opaque taken in their singularity, but that in combination, with enough heat, moisture, and spin from the ocean-born pinwheel, produce a solstice. This is not a solstice defined by the sun's proximity to the equator or a maximum or minimum point in the hours of daylight; rather, it is a solstice marking the sky's ability to start fires that are self-sustaining.

Undeniably, winter thunderstorms do occur. Disparate air masses collide, causing instability and cloud buildup. These storms are what I picture filling the backdrops of old black-and-white horror movies, and they conjure up images of Young Goodman Brown shuffling down dark, tangled paths in Nathaniel Hawthorne's Salem. But in the sage-covered high lava plains of Oregon, what inaugurates spring typically spins out of the south, born over water to the southwest.

In this ancient springtime ritual, the definitions of earth, fire, and water intertwine amidst a convoluted dance of cause and effect. This is a mystery fit for the most devout gnostic: Out of water comes fire; a pillar of cloud becomes a pillar of fire; and from whence comes fire also comes its demise. The cycle is recapitulated upward as the pitchy hollow of a juniper tree or the thick, crispy duff beneath a ponderosa pine smolders, sending up wisps of smoke in the wake of a lightning bolt's hot blast. Granted, lightning isn't the same sort of fire that transforms a splintered strip of juniper bark into flickers of flame, though it is more than analogous: Lightning shares some of the properties of fire—heat, light, and the ability to kindle flames. Maybe the ancients really weren't that far wrong to see lightning as true fire.

✦ ✦ ✦

April is the month of infant fire: Prometheus's gift to earth, reborn spring after spring on the high desert, is an amniotic state of affairs in its first gleamings. Any terrestrial fires ignited by the fire from above are quickly aborted at this embryonic stage. Typically, the rotten cores of junipers are filled with punky wood and nearly fossilized rat crap. Once ignited, they'll cradle smoldering embers for a day or two while their exterior bark slowly dries from the spring gully washer that conceived the whole event in the first place. Other trees are likewise seared, but most of these bear only scars of stripped bark spiraling the length of their trunks, indiscreetly revealing their underlying cambium. Sometimes needles and branches lay scattered in shards for some distance around the base of a tree. But, this time of year, rare indeed is the solid tree that holds any fire; fire must remain sheltered if it is to survive in this season, which is, even for humans, potentially hypothermic.

In the juniper that does still ensconce embers several days hence, fire may gain enough momentum to climb its way up through branches, cresting out onto sticky green needles and dull-blue berries. Seldom, this time of year, will the tree become totally engulfed in flame, but it will put up enough smoke to elicit one or two frantic 911 calls: "There's a fire burning at —." Some BLM or U.S. Forest Service employee will usually respond to the conflagration, if for no other reason than to get a head start on hazard pay, however feebly justified. If you're the one responding, and you're lucky, you won't get your fire engine seriously stuck in the mud. For those in the ranks of the so-called seasonal firefighters (who for various reasons commit themselves to temporary employment and inevitable layoffs), these lightning-born harbingers of summer are temptations to drag out black, Vibram-soled Wescos, Whites, or Danners, if not to grease them—for there's still time for that— at least to make sure there is enough tread for another season.

The infant fires of early spring are still inextricably bound to water. Storms remain wet, and the fire at the center of our solar system has yet to cure earthen vegetation that is only now beginning to fully sprout. Beneath the desiccated-looking branches of sage, rabbitbrush, and bitterbrush lies a luxuriant carpet of green—some of it is recognizably new growth from perennial bunch grasses, some is nondescript blades, most likely cheatgrass.

Into this verdant newness, humans introduce fire. A handful of seasonal employees—like us fencers—whose tours of duty just began, directed by an even smaller group of full-time "permanents," cram fire down nature's throat with the assistance of drip torches or pouches of Alumagel (a napalm-like substance that resembles the gelatinized grease that clings to canned hams). According to detailed prescriptions that give environmental parameters for unleashing fire (including temperature, fuel moisture, and relative humidity), these folks go out and light off what concentrations of fuel will burn this time of year, and sometimes that means only the fuel in their torches. They do so under sundry banners, such as fuel loading and fire hazard reduction, or wildlife habitat improvement. Sometimes objectives are reached; other times, the foreign fire—delivered from the greasy silver canisters, which look like flaming watering cans—is swallowed up as soon as its petroleum nectar trickles to the ground. If this is the case, spring must be allowed to penetrate a little deeper into summer before burning prescriptions can be filled. However you cut it, fire is just crawling out of its infancy and barely in its adolescence.

To introduce fire into nature, at any time of year, is to recognize the formative role that fire has played in shaping and maintaining the integrity of ecosystems over time. To introduce fire during spring is to attempt to manhandle it, to keep it within bounds, and to direct it where and when we want it, with as little chance of negative impact on humans as possible. Such machinations may be entirely in order. Still, the fires of spring are not the fires of summer, no matter how much diesel and gasoline you use, nor are their results the same. I'm not sure what to make of these matters, but like many important truths, the significance may lie in the details. How the still-

damp carcass of a downed juniper responds to fire when sun-
dried grass is lit off along its sides, how many days before the
summer solstice this occurs, or whether a covey of quail is
nesting in the tree's shelter are all questions whose answers
ripple down through the soil and across the landscape (physically and morally) in unforeseen ways.[1] Only ignorance blurs
the differences between separate fires, seeing charcoal as
charcoal, and fire as fire, no matter when or where they happen. These are matters of science, and they are also matters of
philosophy: What is the value of landscapes or configurations
of plant species? Is nature's fire qualitatively, or only quantita-
tively, different from that unleashed by a human hand? Are
human ends the only ones of ultimate importance in all of
this? Scientific and philosophical, yes. Possibly even religious.

Eventually, in one or two months, infancy gives way to adolescence. Until then, intermittent storms will produce sleeper
fires that may or may not froth up visible smoke. Dispatchers
will keep busy trying to do the politically correct thing with
those 911 calls, and those few government employees kept
around during the off-season will make obligatory trips into
the intermountain forests or high desert to dig moats around
dying embers, because that's policy. Countless fires will run an
accelerated life cycle from birth to death without ever being
detected. Like the floral life that is just budding, fires are
barely gaining a foothold. Even a campfire is no easy task to
keep going.

2

Axis Mundi

The Fire. The odor of burning juniper is the
sweetest fragrance on the face of the earth,
in my honest judgment; I doubt if all the smoking
censers of Dante's paradise could equal it.
One breath of juniper smoke, like the perfume
of sagebrush after rain, evokes in magical
catalysis, like certain music, the space and
light and clarity and piercing strangeness of
the American West. Long may it burn.

—EDWARD ABBEY, *Desert Solitaire*

For a number of years while I was in college, several friends and
I made an annual spring pilgrimage into Oregon's Deschutes
River canyon, just upstream from the slack waters of Lake Billy
Chinook. There, we coaxed native redside trout out of swirling
eddies and gazed into the crackling fire that we built year after
year in the same basalt ring. The starting and tending of such
fires were always important rituals, no less important than the
finished product of light and sparks and smoke issuing forth of
their own accord between the rocks. Days were for fishing; nights

were for fires, and night would find us sitting or standing around the flames. Though the details of individual trips have blended through time, their progression remains palpably clear and distinct.

We secure our rig on a sage-covered plateau high above the river. Snowcapped volcanoes line the western horizon—Hood, Jefferson, Three-Fingered Jack, Washington, the Three Sisters, Broken Top, Bachelor—familiar places with names on maps. We then angle to the northwest, down a quarter-mile trail to a riffled strip of blue at the canyon bottom, a place familiar to some but obscure enough, and subtle enough, to evade U.S. Geological Survey namers of things. Golden eagles convect overhead and perch in twig-fortified alcoves amidst the uppermost rims. Below us, kingfishers sit motionless on alder branches overhanging the river, staring into the blue-green ribbon of cascading liquid.

Bailing off the flat, we drop through bronze and black colonnades of lichen-encrusted basalt, then through bands of cream and gray sandstonelike conglomerate—a petrified layer cake. The stratum beneath the basalt is a sort of dehydrated aquarium: The nearly 100 percent slope of its angled face displays ancient clamshells and white, pumicelike scree mixed with little fragments that glisten like mother-of-pearl. Though we're still a good four hundred feet above the river, several dozen million years ago there was obviously water in abundance. Eventually, the prehistoric silt was capped by a fiery mix of molten lava, which oozed and smeared out across what is now northern Central Oregon. In time, the lava cooled and hardened into the sharp geometric pillars and columns that to the trained eye belie its lineage. But tonight, still six chains (396 feet) below our sinuous trail, there will be fire.

Each of our steps carries us several feet down the slope, and back through geologic time; only upon reaching the canyon floor do we catch up to the frothy ripples of the present, and the present is fleeting. The sandy pumice is loose, and once a boot breaks through the first inch, the soil is damp and dark. The hike downhill is truly an exercise in balance and surefootedness, skills that not a few lacerations on appendages and fishing rods suggest we sometimes lack. It doesn't help that on some of these trips we each lug enough gear to outfit a string of Tibetan Sherpas. (For a time, my only sleeping bag is a canvas and flannel brute—the kind with pictures of quail and pheasants printed on its quilted interior, the kind that compacts into a roll three feet in diameter.) As if to mock us, some sizable boulders also defy the slope and the pull of gravity, perching on thin-waisted spires of tightly packed sediment.

Here and there, scattered in with the scree and sage on the slope, grow bright-red Indian paintbrush and dark-blue lupine. They signal that, at least on the west and southwest slopes, spring has arrived.

In this landscape of angularity, our camp is one of the few level places. The east side of camp is bordered by a boulder patch that rises to the sage-covered plateau several hundred feet above. Immediately to the west are the willow- and alder-lined river and a few stately ponderosa that were just taking root at the time of the American Revolution.

Upon reaching our campsite, it's always a toss-up whether to first unsheathe rods for a sample cast or two, or pitch tents in preparation for nightfall. The risk of immediately breaking out tackle and testing the waters lies in the oft-experienced reality of setting up tents in the dark on what had appeared, at first glance in the daylight, to be perfectly flat ground. Only later, when snugly ensconced in your sleeping bag, do you discover that every rock, pine cone, stick, molehill, or gopher hole in the county has materialized beneath your tent. Here lies the hazard of sample casts— casts that have a habit of elongating into hours, "One more cast" trailing off ad infinitum into "Another," "Just one more," "This is really the last."And, besides the difficulty of setting up camp by Braille, it's always hard to find good wood in the dark.

So we set up tents and we search for wood in the midday sun. We hope that spring isn't burrowed too deeply in the limbs and sticks and brush that will constitute our evening firewood, for we have only so many matches, squares of toilet paper, and white gas to help sustain feeble combustion. We need every advantage we can get, for it's not easy to lure heat out of hypothermic growth rings.

Unlike the fires of summer, when the barely visible flicker of a wooden match touched to dry seedpods can set in motion a snapping wave of flames, the fires of early spring must be tempted and enticed into the open. No haphazard arrangement of twigs will suffice as a receptive fuel bed to a lit match. Large branches must be laid horizontally in the fire ring to serve as fulcrums upon which other woody debris is piled. Of equal effect, the edge of the rock rim works well to support the little lean-to that must delicately shelter the first breaths of fire, for fire does breathe: Life-sustaining oxygen must saturate the just ignited tinder, since it is as critical for the sustenance of inanimate combustion as it is for the cellular combustion at work in animate life. Next, under this shelter—either against rock or between horizontal branches— you must lightly stack tinder, usually composed of needles and pitchy slivers, but sometimes dead grass or bits of bark.

The bases of large, old juniper and pine are the best places to prospect for dry tinder, the trees' thick torsos having shunted away the majority of winter and spring precipitation from the rusty needles and yellow filaments of grass that now encircle their trunks. Also, the spindly frames of these needles and grass blades dry out much quicker than the dense, waterlogged wood that lies scattered in the open.

Finally, having gathered up these tidbits of dryness and bedded them down under your shelter, you massage some heat into them. If the bed ignites, well and good, but you had better have more tinder on hand. How many times I've turned my back, distracted by some other urgency, only to glance again at the fire pit to see a small pile of clinically dead, smoldering ash nestled beneath unscathed twigs. After releasing a salvo of expletives, I begin again. If, on the other hand, you can find a few slivers of pitchy wood or papery strips of juniper bark to add to the newborn fire, enough heat may build to allow you to add larger sticks. If momentum wanes, you frantically root around for a few handfuls of cheat, more needles, or whatever else is handy and dry to heap on the sickly flames. These last-ditch maneuvers may very well smother the tiny blaze or strangle it to the point of casting a shadow of ambivalence over its fate—smoke eerily boiling out between intertwined needles and grass, creating the ambiance dry ice produces in a stage play.

But often, just when the pall of death blankets the fire pit, in a flash, the white smoke flowing off the duff transforms into jerking and twisting fingers of yellow and red. Like a trout rising to a bobbing mayfly, the fire takes the bait. To keep it hooked, though, it must be played with finesse, feeding it only as much fuel as it can consume.

Even though my friends and I conceive and give birth to these first meager flames that later in the evening metamorphose into a three-foot-wide conflagration of sage, alder, and juniper, the fire is very nearly free burning. This is evidenced in the continual game of musical chairs we play around the fire ring: As smoke drifts this way and that, we are now the ones responding to the fire. We control where we place the fire; we control how we start it; we control what we feed it; however, our control is incomplete, for the wind ranges freely amidst the flames, providing valuable oxygen to refresh the embers and to sting the eyes of anyone downwind from the drifting smoke. And to the extent that even this fire is wild—a happening that we instigate, then nudge, but for the most part watch unfold of its own accord—it is a symbol of our tenuous relationship to the forces of nature, and a harbinger of the fires of summer.

Could it be that the fire is a mere catalyst for the weighty musing and philosophizing that bind us to its rocky perimeter late into the night, as if it is a center of gravity that forces us to look either into its throbbing glow or into the faces of those encircling it? *Axis mundi?* The navel of the universe? Discussions might begin with small talk, but in the radiance of the embers and flames, dross quickly burns away, replaced by what really matters: fly patterns, the ethics of letting suckers rot in the brush, arguments for and against God's existence, the meaning of life, and, always, whether it will be a good fire season.

On the other hand, maybe the fire itself is more intimately bound to the subject matter of our conversation than simply drawing us into an isolated and focused circle of contemplation. We tend and watch, watch and tend our fire. We also watch and tend our conversation. I suspect that what we find valuable in our communion around the flames is just that—communion *and* fire. And maybe what we experience on such dark nights in the face of the flames becomes real only because of the flames.

The rocks built up around our furnace are smooth, weathered by countless years of water and sediment and other rocks grinding their surfaces. Quite to the contrary, the fire between these stones is capable of rapidly transforming whatever contacts it, be it wood, flesh, or stone, sometimes even breaking open rocks that contain hidden seeps of moisture, expanding and fracturing them, blasting sparks and rocky shrapnel out of the pit in a burst of sound and light.

Depending on the vigor of our little blaze, light may spread for some distance beyond our immediate circle, illuminating tree branches and nondescript brush and grass, never bright enough to keep our attention for long. Beyond that, there is darkness, sometimes phosphoresced by moonlight, always difficult to fix one's gaze on with pupils dilated by the fire. Finally, if one looks up and strains to see, there are stars, but they are hard to distinguish from the nebula of convecting sparks. One must step away from the fire to see the stars clearly.

Much to our horror (though often evoking laughter after the fact), we sometimes wake to discover blackened ground radiating away from the fire ring, both bespeaking hesitance (negligence?) on our part to extinguish the fire before going to bed, and testifying to the fire's stubborn hunger and will to find fuel. Released from our poking, prodding, and controlling gaze, it squeezed through passages in the ring and found limited life in the sparse ground litter of needles and matted grass.

I learned much from those fires beneath the rimrock.

✦ ✦ ✦

Fire can be dramatic, which is one reason that humans have found it so instrumentally useful. But there is more that utility does not even approach. Even a child knows that a fire can cook hot dogs and marshmallows or incinerate them into hardened chunks of jet-black creosote. This is important, but it is not the whole story. Besides the value of fire's ability to transform matter, there is a value intrinsic to it in only the watching. I'm not even sure if the watching can rightly be called pleasure, though it is valued. The metamorphosis that we witnessed as our campfire developed from a lazy smolder in juniper bark and pine needles to the concentrated heat that would cause pockets of pitch to boil on chunks of gnarled and knotty limb wood evidenced something that wind and water are hard-pressed to do: proclaim a secret about the world—life is fire. Fire's window into reality provides a view into the *potential* of the world. Within the growth rings of dried wood lies the ability to put forth heat and light, to become brittle in its own pulsating glow, and, after giving up most of its weight and mass, to remain as a still-warm but cooling pile of ashy flakes and blackened charcoal bits. To see this, to feel and smell that this is a natural part of the world—punctuating the drone of sameness with radical transformation—satisfies a need to know about the world and dazzles us with a knowledge that all is not what it sometimes seems.

In this informed seeing there is beauty and revelation. Colors and patterns of light rarely seen and incapable of being fashioned with oil or chalk burst forth in entirely unanticipated ways. Crisp pine needles snap vigorously, as do slivers of pitchy wood; sun-yellowed grass burns quickly and then, as if in slow motion, floats upward in the form of delicate white ash. Being in the presence of such variegated—yet usually hidden—natural processes is meaningful knowing, and knowing combined with meaning is true seeing. Without a doubt, heat and light and flame, which are qualities of fire, richly symbolize and mirror the interior dimensions of the human soul—energy, power, life. Fire is also beautiful and revelatory of what lies beneath superficialities. Fire evokes amazement and begs recognition of the mystery in things.

That half-rotted alder branches can sustain themselves in combustion or that green willow switches will smoke and wither in a flame are amazing facts—amazing when I'm watching, and just as amazing when my back is turned to the fire pit and my eyes are closed from the sharp sting of smoke. There is the flavor of objective reality to it all; the secrets one beholds in this happening transcend metaphor and resonate with the truth that one is witnessing something very basic, sturdy, and foundational.

Watching a fire—contained in a hearth, surrounded by a rock ring, or snaking its way through mountain canyons—one does not perceive that anything like a chemical reaction is occurring within the superheated gases sublimated off chunks of woody fuel. Rather, one sees reds, blues, yellows, each undulating and spiraling and then dissipating into whites and grays; one feels the warmth of this event process, sometimes comfortable, sometimes searing; one smells a fragrant odor of burning pine, juniper, and oak, or the repulsive odor of burning dung, or the urine of some whippersnapper who let go into the fire pit; or one frantically inhales the choking combination of smoke and heat before unconsciousness sets in and death soon follows.

◆ ◆ ◆

So in May, along the Deschutes River, next to ancient, cat-faced pines trimmed with nodules of amber pitch, and along water that not too many years ago provided safe passage for wayfaring salmon and steelhead to their nuptial chambers, we built fires. If for no other reason, we did this because spring is the time to tend and watch fires, and it's chilly in the evenings. Unlike the months that preceded this seasonal ritual, this was also a time to *hope*— looking toward the ever escalating role that the hot, ethereal stuff seething in our rock ring would play in the months to come. No season is without hope for something. If we're sick of rain, we hope for sun; if we're tired of slush, we want dry cold; if we're sweltering, we long for the chill of autumn. But for many of those who will greet fire several months hence with pulaskis and shovels and wet gunnysacks, the hope of spring signals a time to prepare. For those of us who kindled these little blazes along the Deschutes and who would later engage fire more actively, these nascent embers were sacraments of hope and preparation, necessary to value fire on a grander scale. The springtime of our lives paralleled the springtime of fire, both serving as signposts, and both subsisting in the light of watchfulness, hope, and preparation.

◆ ◆ ◆

For all his obscurity, I think Heraclitus of Ephesus was on to something. The "reversals of fire" do, indeed, involve sea—even if sea be spun inland to grow cauliflower suspended in the sky. Though I'm not sure what he meant by half of the sea being earth, surely a part of the sea does involve lightning. Out of spring lightning, ground fires break forth, timidly at first, and with great effort combined with a little luck.

Summer

3
Converging

If I should only give a few pulls at the parish bell-rope,
as for a fire, that is, without setting the bell,
there is hardly a man on his farm in the outskirts of
Concord, notwithstanding that press of engagements
which was his excuse so many times this morning,
nor a boy, nor a woman, I might almost say,
but would forsake all and follow that sound,
not mainly to save property from the flames, but,
if we will confess the truth, much more to see it burn,
since burn it must, and we, be it known,
did not set it on fire,—or to see it put out,
and have a hand in it, if that is done as handsomely;
yes, even if it were the parish church itself.

—HENRY DAVID THOREAU, *Walden*

The hopes and preparations of early spring gain intensity as May and June draw nigh. In the rumble of distant thunder, a parish bell sounds across the West, and a mass exodus begins from a legion of far-flung locations, from Corvallis to New Haven, Durham to Boise, Anchorage to Santa Barbara, and Seattle to Boston. All the travelers converge with the common goal of meeting fire along margins separating green from black. Like Thoreau's Concord neighbors, these travelers are not emigrating mainly to see the summer's wildfires put out. Rather, to have a hand in such a task, to look deeply into nature's crimson face in July and August, and, at times, to do so handsomely, is a large part of what draws— and keeps drawing—the modern emigrants down the Oregon and Santa Fe Trails, the Applegate and Meek Cutoffs, and to parts beyond. If spring is dominated by watchfulness and expectancy, then summer is surely the time of active engagement.

Yes, there's money to be made in engaging the fires of summer, and, for many a college student who measures nutrition in Ramen noodle packages per dollar, the success of a fire season is measured in hours of overtime. Yet, I suspect that even those who say they are in it just for the money are also—though inarticulately—spurred on by deeper motivations. Anyone who has listened to a firefighter or ex-firefighter reel off stories of exploits and experiences of hot, smoky July and August days knows that there is more to fighting fire than greenbacks. Some identify the locus of attraction in the chivalrous category of risk; others reduce it to physiology—adrenaline rush. Undoubtedly, both are true. But I think there is more—a residual more that, listened to, reverberates deeply into, and then back out of, the human soul.

Because fire embodies more than mere employment value, engaging fire is never just a temporary job, as if temporary and seasonal jobs should be juxtaposed against "real" jobs. (Is fighting fire, then, fake?) I've heard many people make just such contrasts, and while there is little to commend in the usual drawbacks of temporary/seasonal work—meager benefits, little long-term stability, bottom rung on the bureaucratic ladder (which may be a blessing)— the work is nonetheless real, and I, for one, wouldn't trade it for many a so-called real job with all its real, envied embellishments and its real ulcers.

By the middle of June, as the great migration is underway, a few of the emigrants have already begun playing with fire. Some have gone through the perfunctory motions of extinguishing single-tree blazes—blazes that are frequently surrounded by waterlogged playa, saturated long before initial attack resources ever arrive to add their doses of water to the scene. Other emigrants have been their own lightning bolts, dribbling prescriptions of fire from drip torches into timber-thinning units and slash piles.

As spring trails away into the imminent expectation of summer, thunderstorms shake off their aquatic ancestry and take on more terrestrial qualities. Like so many caddis wriggling free of their amniotic streams and lakes to live out their days in the realm of air and land, the storms of late spring become amphibious. And with the drying of the buildups and the lengthening of the days comes a drying of the land, and a maturing of fire.

The fire of spring and the spring of fire are luminaries leading up to the conflagrations of summer. In some years, there are no perceptible differences between these seasons, as summer unfurls in a lengthy attenuation of spring. The heat never comes, and precipitation is abundant. Springlike storms roll in intermittently, mere punctuation in layer after layer of stratus clouds poured in by the jet stream, a meteorological run-on sentence. Eventually September comes, the lightning becomes rare, and with its absence evaporates any hope of summer. Such a year is like a bad dream for those who hear the parish bell, a summer that never was. Statistically, these summers should be expected; nevertheless, they are never *hoped* for.

There is a parallel, though not an exact correspondence, between the summer solstice and the summer of fires. The greatly attenuated daylight of the northern latitudes that peaks in late June allows ever more solar fire to penetrate the earth's thin film of life and air. How this affects terrestrial fire is a subject of generality—that is, a subject peppered with exceptions, degrees, and shades of variation. In drought years, or in places such as Alaska's interior, fires will blossom into adulthood at or before the longest day of the year, evidencing the malleable nature of fire's maturation. Some fuel types cloak resins and waxes that stubbornly resist the death knell of high relative humidity. In years with heavy snowpack, alpine cirques in the Rockies may not see smoke until late August. Yet an occasional wind-driven fire might ramble across thousands of acres of chilled Great Plains prairie during January. Is it an orphan carryover from summer, or a premature eruption of the season to come? Generality.

For those who travel to the margins of fire, "summer" is likewise a term of generality. In one sense, summer embraces the totality of one's tour of duty committed to engaging fire. In another sense, the season of summer employment begins not in spring's demise, but in its fruition, and in an ironic mix of death and life. Feathery-seeded cheat, once brilliant green, is by late spring or early summer in various moribund stages of purple and yellow. By July, summer is set firmly into its stems, and, once its vegetative life has seeped out, the life of mature fire can reinvigorate it with crimson waves of flame. Other plants will continue to grow and bloom throughout summer, even if they become slightly dehydrated. And when exactly—or, better yet, generally—this happens will vary from region to region and from year to year.

For instance, New England is sometimes frugal in its generation of cumulonimbus in March and April. In the land of the nor'easter, even when thunderstorms do build up during this leafless and gray period, their classic anvil shape is usually shrouded in dark masses of undifferentiated clouds. These are better than nothing for someone searching for signs of a new season of fire in a land far removed from sage, ponderosa, and rimrock, but they are only half-siblings of storms that will bring fire to earth and earthen fire back up into the Western skies. Occasionally, a lone emigrant firefighter driving west from New England or below the Mason-Dixon Line will encounter the first recognizable relatives of Western storms silhouetted against Midwestern corn stubble and whitewashed silos—or maybe at Chimney Rock, Nebraska, where travelers over a century ago also marked a significant point in their journey. But wherever cumulonimbus are encountered, and however grotesquely they resemble the clear, dry margins that will roll in from the Pacific, they are still reminders that a new season is on its way, and nature's ability to forge embers has not deserted the earth. Whether or not the emigrants hear the perennial parish bell is of no account, for with the knowledge that mid-June and the fires of summer (and only secondarily a bevy of federal and state agencies) require their services, they turn into the current and head west.

For those participating in the rites of summer firefighting who live in the West, by May, New Year's has undoubtedly arrived, and, with it, expectation and preparation. Opening up a ditty bag, rucksack, or a box labeled "fire gear" is like opening a time capsule. Black leather boots that have sat idle for months are dragged out and baptized with fresh layers of grease or neat's-foot oil. Green, fire-resistant Nomex pants are discovered in duffel bags hibernating in back corners of closets and greeted with half-sincere looks of surprise: "Hmm . . . I thought I turned those in last fall." (The reality is that one grows sick of annually hemming agency-issue pants cut for Goliath.) Yellow fire shirts are found—some still stiff with dried sweat and stained with chain-saw bar oil or red retardant speckles. Often in their pockets nestle sundry treasures: a couple of dirt- and wax-caked earplugs, a tattered strip of fluorescent orange ribbon, and maybe a few pine needles and juniper berries. These are all reminders of a past season. Other icons inhabit musty grottos in one's home: a handful of dead AA batteries, a stack of T-shirts reserved for summer months (like medieval escutcheons on shirts of chain mail, on them are emblazoned the names of fires or fire organizations that mark one's fire clan and history), and maybe a folded shift plan from some exotic fire assignment.

Sometime in mid-June, the first leg of the mass exodus concludes. Firefighters and wanna-be firefighters of all ranks, backgrounds, and specializations arrive at their penultimate destinations—to regroup, retrain, and re-equip. They converge on such disparate towns as Winthrop, Battle Mountain, Prineville, Fairbanks, Missoula, McCall, and Las Cruces. They've come from a thousand different starting points, and from their penultimate bases will jump off to a thousand other destinations, be they anchor points on wildfires or intermediate staging areas such as remote guard stations or lookout towers. The margins of my fire home include history-laden names such as Maupin, Paulina, Ochoco, Hampton, Aldrich Mountain, and Tower Point. They can be pinpointed on maps and hemmed in by township, range, and section lines; the earth for which the names are signifiers can be palpated; and the air coursing through their valleys or over their ridgelines can be breathed in, as it was by the pioneers and American Indians for whom they are named. The names also stand for a world, albeit a world whose margins butt up against other possible, and actual, worlds of fire. They are geographic and temporal markers whose hinterlands fan out until pressed back by rivers, gorges, and mountain ranges that define the margins of other firefighters' home worlds.

✦ ✦ ✦

My official journey into the realm of wildland fire began on an unseasonably warm mid-June day, in a stuffy little classroom in Central Oregon Community College's Ponderosa Hall: Fire Guard School. Several weeks earlier, the assistant fire management officer for the Prineville District of the BLM had called, offering me a fire-engine crew-member job. I accepted. But rather than putting on basic fire training for me and the other rookie they were hiring that year, they decided to ship us over to the Forest Service Guard School at the community college in Bend. At the time, I wasn't too sure what I was going to be guarding—and I'm still not too sure.

Wildland fire was not completely foreign to me. I'd fought a couple brushfires as a member of the Crooked River Ranch Volunteer Fire Department. In actuality, it was more like I'd flailed along their margins, throwing dirt, spraying water, and trying not to get burned. I thought I was a seasoned expert, even if wildfire was something that most of us in the department learned on the job, and the jobs weren't that common. I learned that while fighting brushfires during the blazing heat of summer, there had to be a happy medium between wearing shorts, tennis shoes, and a tank top and being fully decked out in flannel-lined turnouts—there's a golden mean between severe burns and heat exhaustion. What I didn't learn was that all the water in the world wouldn't mop up a fire without digging and cold trailing (that is, feeling with your bare hands for hidden heat), that there was a thing called the wildland fire environment (weather, fuel, and topography), and that wet burlap can extinguish burning grass. That would come later. In the meantime, I enjoyed the excitement of tone outs—my pager screeching tones and beeps alerting all volunteers to head to the fire hall—and the feeling that I was doing something that mattered.

My introduction to federal wildland firefighting actually began while I was working as a volunteer riparian inventory technician with the U.S. Forest Service, the summer before I attended guard school. While I was flipping through dichotomous keys to identify renegade willow species, measuring tree diameters, and trying to make ends meet on my meager per diem salary, a close friend of mine, Robert, told stories of fire and overtime during his first season with the BLM.

The next spring my application was off to Denver, and, in June,

I was sitting in that stifling-hot guard-school room with people I didn't know who probably knew more about fire than I did, or so I thought. In fact, just about everyone present was new to fire; many of my peers in the class were regular Forest Service employees whose official duties had nothing at all to do with fighting fire.

There were four BLM employees in the class: me and J. J., both on the fire crew; Rachel, in administration; and Berry, in recreation. In our unfamiliarity we had two things in common—we were all required to be in this training if we wanted to fight fire, and our paychecks were printed in the same place. So we clustered up and shared our anxiety.

The week-long itinerary included such cryptic notations as s-190, Introduction to Wildland Fire; s-130, Basic Firefighter; Water Handling; Fire Shelters; Group Fire. Besides the sweaty, stale, burnt coffee–smelling air in the room, there was also the qualitative air of importance about the mission we were embarking on. Folks clad in light-green uniform shirts—with U.S. Forest Service patches on their shoulders—gave pep talks. Blank flip charts stood on either side of an overhead projector at the front of the room. I hoped that the instructors were going to write on these; secretly, I feared students might actually have to venture forth to divulge their ignorance and inability to spell correctly in the presence of the entire group.

A movie projector sat in the middle of the dusky room. Enough light filtered in through the window blinds to allow note taking, if anyone felt so inclined. Someone turned the projector on. An ill-focused square of light appeared on the screen, alive with shadowed, wriggling lint particles, like so many amoebae under a microscope. Then flashed the words:

UNITED STATES DEPARTMENT OF AGRICULTURE

FOREST SERVICE

presents

FIRE WEATHER

The film's narrator was a balding, middle-aged, professor-looking character. He began by stating that the main weather concerns for the firefighter are wind, lightning, humidity, and fuel moisture. I scrambled to take notes. By the look and sound of the film, it was probably produced sometime in the 1960s, the kind of film students love to mock. But it made fighting forest fires seem important. The film concluded with a case study of the Sleeping Child Fire in the Bitterroot National Forest—a fire that did anything but sleep, being whipped into a conflagration by something called a "cold front." Not a few heads were nodding as the room began to bake and the well-manicured professor droned on. The final strain of symphonic background music came to a crescendo—

THE END

"Take a ten-minute break and be ready to start again at 1300," one of the green-shirted instructors bellowed.

How the hell are we supposed to keep straight foehn winds (pronounced "phone" or "fern"?), cold fronts, burnouts and backfires, azimuths and declinations, and relearn telling time all in the same week? I thought to myself. Somehow we did.

Our field training was cut short by about a day. It had been a dry spring, and lightning-caused fires were already breaking out on the Fremont National Forest to our south. Firefighters were needed, we were told. The cadre of green-clad instructors hit the highlights of compass use and fire shelter deployment, and we went out into the woods, where we broke into small groups to extinguish practice fires, each a mere two chains (a chain being 66 feet) around. Why can't firefighters talk like everyone else does, in good old feet and yards? So I wondered then, and I still wonder.

We took a final exam and filled out a course evaluation. Then we were told to return to our home units and prepare for our first fire assignments. Many of my peers were soon dispatched to the pumice-soil ponderosa and lodgepole forests of the Fremont. J. J. and I grabbed an engine in Prineville and headed north to a reported fire along the breaks of a river called the John Day, a fire burning below the still-green margins of winter wheat.

◆　◆　◆

Those intent on directly engaging wildfire do so in many ways—as engine crew personnel, smoke jumpers, lookouts, helitack (helicopter attack crews), hotshots (twenty-person organized crews), hand crews (ground crews with hand tools), or pilots. Obliquely engaged are dispatchers and a host of miscellaneous other support personnel. Each group exemplifies a sort of guild, possessing a craft by which it contributes to the economy of fire. By "economy," I don't mean the trading of services for cash. Rather, I mean a way of being and encountering the world of fire. This includes starting fire, putting fire out, and organizing one's time and day planner to allow for this engagement. Most firefighters will belong to federal or state agencies. Some who inhabit the fringes of human settlement might join the fiery rituals of summer in the ranks of city and rural fire departments, as I did for a time. Some enter through the free market.

Through this laissez-faire entrance into fire's economy comes the private contractor. From the standpoint of those who contract out their services for fire suppression, fire is a problem to be solved by services traded as commodities. Theoretically, the contractor supplies both the skill and the equipment to the public at a price more competitive than the government's, to the benefit of both the public and private sectors. Maybe this is sometimes the case, and for many fire suppression services this may be the most cost efficient and sensible way to operate. Moreover, one needn't be terribly perceptive to realize the extent to which Uncle Sam is wedded to waste. Yet there is something obscene about engaging fire if it is reduced to profits and cost-benefit analysis.

Engaging the primordial flames, which rekindle year after year as spring buds into summer, is meaningful, and meaningful in the richest sense of the word. Of course, virtually any work that merits remuneration may accrue significance beyond contributing to a bank account. It may even be human nature to ennoble what may be ignoble, to give meaning and value to what is confused, trite, or flat meaningless. But the illumination fire provides is more than visual, and to speak as if all that fire illumines is the dance of dollar signs in the flash of flames is to treat engaging fire as a mercenary activity.

Fighting fire may include being paid for one's time; it also means contributing to a common cause that involves harmonizing our endeavors with the purging force of nature's flames. This, too, transcends the mercenary. At one time in the history of wildland fire suppression, this meant controlling every fire by 10:00 A.M. the morning following its report.[1] Today, this practice has (in theory) been replaced by more nuanced and ecologically informed policies. Since the fires of forest and range by definition require the destruction and consumption of life (or that which was once alive—be it old-growth Douglas fir or the rust-colored pine needles carpeting the forest floor), engaging fire evokes ethical questions. Is extinguishing *this* blaze something *I* ought to be doing? Is it okay that I enjoy seeing flames? Is excitement in the face of a fire a motivation to keep facing it? Would the world really be a better place without fire? Whatever the answers to these questions (and their pursuit is important), because we live in a world punctuated by fire, and punctuated in such a way that not all fires will always and everywhere be tolerated, this human-nature, or nature-human, economy of fire depends on those who measure seasons in the buildup of cumulus and the driving rains of fall. And this includes all firefighting guilds.

Each guild plays an important role, and each has its own language built around the transition from apprentice to journeyman.

Some individuals like to claim not only that this transition is internal to a specific guild, but also that some ways of engaging fire are inherently *better* than others. This I strongly doubt.

I'll waste no time extolling the purported virtues of one group over another, as if glitzy epithets could elevate some of the fire community to sainthood. Some groups may, on the average, be better trained and in better physical condition than others, but this, like the definition of "summer," is a statement of generality. Hotshots as "shock troops"? Smoke jumpers as "paratroopers"? These imports from military jargon are nothing new to the fire service and to some extent may be valid, reflecting analogous organizational structure, chain of command, or function. For the most part, however, such comparisons turn on the purported heroic and admirable status of their military cousins. The comparisons also presuppose that there is an enemy to be fought— a highly contentious claim when it comes to fire.

Primarily, these different guilds reflect different modes of transportation to the fire, and any method of travel that smacks of danger (such as whirling blades ready to slice off carelessly bobbing heads) will, to the popular culture, seem romantic. Airborne travel, which severs our earthen umbilical cord, easily lends itself to a feeling of adventure and raw experience. Such attraction is to be expected, for attraction often lies in difference, and in a world where gravity is the norm, to defy it and approach fire from above can provide an exciting, fresh perspective on terra firma.

Furthermore, the medium through which one approaches a wildfire's perimeter closely parallels the fire environment itself. For instance, no wildfire exists without being affected by weather: a gentle breeze that tickles flames creeping along the ground; the gale-force winds trailing the edge of a cold front that whip a ground fire up into timbered crowns; the convective uplifting of moist air that transforms lazy chunks of cauliflower into massive anvils; or moisture either invading or dissipating from the air, in turn either drawing dampness out of grass and logs and trees or bequeathing traces of spring back into their photosynthetic husks. Into this vertical medium fly aircraft loaded with firefighters who angle down to the ground as smoke columns rise into the sky.

Fire also requires fuel, and wildland fuel by definition adheres to some sort of topography, be it a gentle, savanna-like plain of Idaho fescue, or severely pitched, darkly timbered canyon walls. On this horizontal plane come firefighters on foot or by vehicle, wheeled, four-footed, or floating.

Either way—from the sky or along the ground—fire is approached in its native habitat, and by entering its environment vertically or horizontally one lives a little deeper in fire's world. To approach fire in its environment is to push aside mediations, and without protective cover, there is always risk. Such exposed living may be more critical to the experience of a truly good life than modernity suggests; this risk, though, need not be reckless. Life is riddled with a wide array of superficialities and mediations. Food is cooked on electric ranges, and homes are heated by warm air forced out of rectangular floor vents; concrete and asphalt separate feet from humus, loam, clay, sand, gravel, roots, and worms; and computer technology stands ever more firmly between one's mind, other minds, and the world. Undoubtedly, part of today's attraction to outdoor avocations that contain elements of risk (like rock climbing, mountaineering, river running, skydiving, etc.) is that they provide entry to the world a little less encumbered by the trappings of our contemporary culture and its sometimes superficial veneer of safety.

Some people talk as if risk and struggle are divine curses on humanity, and modern conveniences are ways to cope with, and war against, that primordially spawned fallenness. I, however, tend to think that there were always thistles in Eden, thistles that dried out and caught fire, that were a bit of a nuisance at times, but whose prickly reality refused to blunt life.

Of course, some ways of entering the fire environment are less mediated than others, walking under your own power without mechanical aids being the least mediated of all. Again, as far as comparing guilds, some ways of traveling do require special training, skills, and physical abilities. Each method of traveling is unique, but all are inconsequential to what is required of a firefighter once a wildfire's expanding perimeter is reached. Whether descending through air or traveling at breakneck speed cross-country, when the fire is met at its combusting edge, all firefighters become terrestrial: Dirt is moved, water is squirted, and bare hands sift through ashes as one crawls along the ground searching out tidbits of lingering heat amidst dirt and rock and duff. As critical and ostensibly glamorous as aerial drops of water and retardant are to modern fire suppression, fires are ultimately corralled (if they allow it) on the ground.

Having said this, though, the (self-)aggrandizement of certain guilds and specializations—by themselves or the public—is nothing new and should definitely be no surprise when fire is involved. People of all professions tend to latch onto their little bailiwick of authority and skill and wear its epaulets and accoutrements in and out of season. For instance, the volunteer firefighter with the local rural fire-protection district proudly struts about with a pager affixed to his or her belt (which was much more impressive before today's pager omnipresence), regardless of being beyond communication range. Then there are those addicted to Western wear, whether or not they own a horse and whether or not the form is truly functional. How many a volunteer firefighter, farmer, or rancher I've seen battling a wildfire in smooth, leather-soled cowboy boots, herring boning his way up the slick bunchgrass slopes like some out-of-season cross-country skier.

Even in the dead of winter, some wildland firefighters still sport their tools of the trade: calf-high leather boots, knives or penlights strapped to belts that are fastened with agency buckles, and baseball caps embossed with agency logos. During fire season, green fire pants worn on and off duty, two-way radios swinging from belts, or flight suits swaggering around local watering holes all carry a sense of importance and connectedness, deserved or not. Sometimes these trappings reflect job-related necessities (even if overlain with pride); other times, they merely display inflated egos.

In terms of show and intensity, the fires of summer also exude vainglory, and until the first autumn rains soak the earth, they show little interest in humility. To speak about fire in this way is to trade in metaphor and figure, which is fine as long as it is recognized as such, but dangerous if allowed to obscure the hot reality that fire is. Fire is unconscious: If fire be enemy or lover, friend or foe, it is one without conscious will or intention. That fire needs fuel, oxygen, and heat found amidst air and ground, and that we enter into fire's own economy of survival when we engage it, demonstrates that the full name of *this* happening—fire—must really be writ fire-nature-humanity.

❖ ❖ ❖

Come July and August, the thick curtains of water connecting cloud bases and mountaintops, so typical of spring storms, have drawn back their moisture, most evaporating by the time it reaches the sun-parched earth. Only virga remains, and the fingers of light that now connect cloud and ground are dry. Late in the day, once the massive pillars of cloud pass, once the tumultuous thunder and spiderwebs of lightning continue north (maybe even up over the Columbia and into Washington, or across the Snake into Idaho), small puffs, and sometimes large plumes, of smoke remain. An epic recurs year after year: A pillar of cloud becomes a pillar of fire. If conditions are right and the latter is well fed and nurtured by heat and wind and cured tinder, the two pillars become one.

To witness this unfolding drama is to be present to the same forces that originally willed fire to earth and earth to humanity. A good share of living meaningfully on this planet involves knowing our relation to what has preceded us and to what will come after us. In the perennial pillars of fire that puff up in the Rockies, Cascades, Sierra Nevada, and Great Basin, we too are connected with the burnt-out cat faces of ponderosa pine, the blackened root wads of wind-fallen aspen, and the coal that trims the margins of road cuts. Fire occurs today as it did a century—even several millions of years—ago. Recognizing this heritage is of cultural and historic value, but it can also embrace the aesthetic, sublime, and spiritual. Even in the act of engaging fire via suppression, our heritage with the flame is not forgotten; in this knowing is respect, and respectful engagement need not entail paranoid fear or psychotic fixation, though it always includes risk.

4

Assigned

Under the heavy season of a burning sun,
Man languishes, his herd wilts, the pine is parched
The cuckoo finds its voice, and chiming in with it
The turtle-dove, the goldfinch.

Zephyr breathes gently but, contested,
The North-wind appears nearly and suddenly:
The shepherd sobs because, uncertain,
He fears the wild squall and its effects:

His weary limbs have no repose, goaded by
His fear of lightning and wild thunder;
While gnats and flies in furious swarms surround him.

Alas his fears prove all too grounded,
Thunder and lightning split the heavens, and hail-stones
Slice the top of the corn and other grains.

—ANTONIO VIVALDI,
"Summer," *The Four Seasons*

July and August are the primary birthing months for lightning and fire along the lower Deschutes and John Day River drainages of North Central Oregon. The Deschutes bubbles up high in the Central Oregon Cascades—angling north through lodgepole, ponderosa, manzanita, and, eventually, sage, juniper, and rimrock. The John Day has its headwaters two hundred miles to the northeast, deep in the Blue Mountains, meandering west through multicolored strata of volcanic ash rich in fossils, jutting north just upstream from the old ferry crossing of Clarno in its final reach toward the Columbia River. The rivers have disparate origins, yet they dump into the Columbia within fifteen miles of each other. In differences there can be commonality: The country that the lower eighty or so miles of these rivers share, as they carve their parallel paths north through layer upon layer of Columbia River basalt, has also been shared with me. This was the province for my firefighting tours of duty with the Prineville District of the BLM.

Much of the high-plateau country between and on each side of the rivers is farmland. Wheat and barley and alfalfa checkerboard the privately owned ridge tops of Wasco, Sherman, and Gilliam Counties. The ridges are dissected by untillable spur canyons and section-size chunks of Sherman big bluegrass. Generally, the big bluegrass indicates land taking a federally subsidized sabbatical from cultivation (compliments of the Conservation Reserve Program); quite a few hundred acres of unsubsidized land also take a break, as light-brown dirt clods lie fallow and naked beneath the withering summer sun and the occasional cloudburst. Down in the canyons, public land dominates. The severe slopes testify to the difficulty homesteaders had eking out a living amidst rock and severely pitched bunchgrass. Back on top, a few irrigation pivots hydrate the soil, but most of the farming is dryland.

Old black-and-white photos line the corridors of the Sherman County Courthouse in Moro. They show scenes of horse-drawn threshers working the dryland wheat fields of early-twentieth-century farms in the county. The landscape looks pretty much the same today, with the exception of modern telltale accoutrements —large tractors and combines that replace the horse teams, a good many miles of paved roads and highways, and the large metal towers that suspend electrical transmission lines connecting Columbia River dams with California lightbulbs.

Like so many regions of North America, these canyon lands have been brushed by both culture and fire. The wildfires that I've experienced and heard about, or those whose blackened remnants I've seen, attest to fire's ubiquity on the Columbia River Plateau. The limping farm communities and scattered, sometimes faint, artifacts of cultivation and habitation in Sherman County speak to the tributaries of culture that have trickled and plowed through its grassy swales and hollows over the centuries and decades. At first there were the Wasco and Tenino tribes, then came the Demoss, Von Borstel, and Reckmann clans.

I've always been intrigued by the little cemeteries scattered along the dusty back roads between the rivers, or adjoining the handful of communities that still support a post office, grocery store, or school. Some of the names on tombstones—even those worn and barely readable—sound familiar, bespeaking several generations that have called, and still call, the plateau and the life of farming home. Sometimes the only clue to a cemetery's existence is a small, rectangular fenced area whose apparent purpose is to corral a few decadent fruit trees, some currant or free-spirited rose bushes, and stirrup-high bunches of thick-stalked, giant wild rye. Whereas many a wood-post fence or long-abandoned homestead has evaporated under the periodic range fires of summer, the tombstones remain. For the living, or for the still-standing homesteads and hermitages of those who once lived on the land, fire erases memory; when it comes to the dead, fire pulls back the grassy cloak shrouding their resting places. Like the traces of charcoal that shade the landscape, the tombstones are links to the past, to people who both used fire and knew an earlier generation of wildfire than the offspring I encounter.

In roaming the swales and canyon breaks in search of smoke, I often felt more familiar with the nearly forgotten vestiges of human habitation than with my present-day neighbors who lived between the rivers. Between looking for lightning, training, maintaining campgrounds, and meeting the margins of fires, there wasn't a whole lot of time to get to know our summer neighbors very well—at least during the span of a single season. But over the course of several years, layer upon layer of small talk invited familiarity with mechanics, post office clerks, grocers, bartenders, volunteer firefighters, and landowners whose spreads habitually hosted fires. Eventually, you remember their names and get a sense of what they think and feel about fire and those who engage it. I credit it a privilege to sit in one of the handful of home-style restaurants (which nearly every little community in the county has) when some local, unaware of my role as a firefighter, feels uninhibited enough to share with some fellow chewer of the fat what he or she thinks about those lazy federal bastards. At such times, it's best just to listen and savor my coffee (weak, burnt, or just damned strong) and stare longingly into the chilled, mirrored dessert case behind the counter.

A farmer from the little community of Kent once told me that family and community dances used to be popular events in Sherman County, but that was before there were so many newcomers. By the look of the abandoned storefronts and dilapidated grain elevators, it's hard to know how well remembrances and perceptions match the contours of history. Maybe it doesn't matter that much. Still, the communities have changed. He might have been suggesting not that there are more people now (for census records show a general depopulating of this land) but that there are more people of a different sort—less attached to the land.[1] Making one's living by charging $1,000 per point on the rack of each bull elk killed on one's land is a far cry from the tilling and threshing and mutual support essential to farming.

In a way, my crew and I were newcomers to this region; in another way—pretentious as it may sound—I felt as deeply rooted to the land between the rivers as those who turned over sod and put down seed for winter wheat. We lived between the Deschutes and John Day, assigned to our station for only three or four months in any given year, but the land and the weather and the sage pressing up against wheat defined the history and geography of some of our souls.

Every year, my BLM fire crew—three crew members and me, the foreman—moved into our fire guard station toward the end of June following two weeks of refresher fire training. Stations similar to ours are found throughout the West, and they serve as testaments to the BLM's upward mobility in the engagement of fire. Few of these facilities will ever appear on the National Register of Historic Places. Many reflect an agency trying desperately to carve out a name for itself in the world of fire—an agency that began on a shoestring budget with a fairly ragtag operation and only slowly legitimized its existence beside the venerable U.S. Forest Service. In our case, the 1960s through the 1990s were a recapitulation of the American dream. From a green military trailer overlooking the little town of Maupin and the Deschutes River, to a single-wide mobile home originally slated for disaster relief and situated on the same site, to a new manufactured home located in Grass Valley atop the grain-covered plateau between the rivers—we moved up.

Even the name "guard station" reflects a philosophy of fire, a carryover from an era in which the lone fire guard served as both lookout and suppression force, charging off on horseback once a smoke was spotted to solitarily scratch a line down to mineral soil, which contains no burnable organic material, around the fire's flaming perimeter. There is something alluring about that image, even if the nobility of the cause for which the guards stood is more ambiguous today than ever: Fire was an evil to combat; man was the guardian of forest and range. This dichotomy is less pronounced today, but between the Deschutes and John Day Rivers—a land devoid of lookout towers—my crew served a similar role, though mostly a symbolic one. The terrain is vast, and views are obscured by gently sloping tablelands and deeply dissected canyons; as a result, most fires are turned in by farmers, ranchers, or aircraft. Yet when the morning fire-weather forecast came across our district two-way radio, telling of a southerly flow aloft bringing in subtropical moisture with the threat of afternoon or evening thunderstorms (and soothsaying predictions of elevated lightning activity levels), midafternoon found us located at obscure little high points—Gordon Ridge, Criterion Summit, Adobe Point, Antelope Gravel Pit—scanning the southern horizon for clouds.

It's hard to explain to the general public that some people eagerly await the coming of lightning, and just as eagerly await the sight of smoke. Undoubtedly, every firefighter has struggled with how to respond to queries of whether or not it will be a "bad" fire year, or to the stone-faced look of the one who asks suggestively, "We're lucky we didn't get very much lightning last month, aren't we?" What to say? Should you be honest? Do you respond to their question with the answer you know they want to hear? Or do you skirt the issue in politically ambiguous terms? All the while, you try to rationalize your answer to yourself, knowing full well that the concerned citizen went away assured that you, too, are a member of the upright cause, at war with fire. So for better or for worse, I did a lot of listening while I drank coffee in the Branding Iron, Round Up, or Oasis Cafes. Whether I thought they were right or wrong, listening to people lament about lightning and federal firefighters (the "Feds") taught me much about people's attitudes toward fires during the summer of fire, even if I acted disinterested as I obliquely studied a fly scaling the lemon meringue pie in the dessert case. Regardless, upon leaving the restaurant, I eagerly looked to the south.

Sometimes the weather reports came to fruition in billowing cumulus that bunched up black and dense down toward Madras, Redmond, or Bend, eventually overtaking our location. With the naked eye, the dark bank of anvils cutting north appeared as a hazy continuation of blue sky. With binoculars, the ambiguous blue revealed itself for the dark buildup that it was. At other times, the day ended with the closest thunderheads a hundred miles to the east, or making steep ascents up the talus slopes of Mount Hood to our west. Even then, the day was not a total loss. There was still anticipation in looking at these distant storms—storms that produced lightning and some fire.

On such evenings, the thick mass of clouds over the upper John Day River appears white from our angle but eventually shades pink as the sun sinks behind the Cascades. Red sky at night, firefighters' delight. Once the stars appear in the inky black sky, small bursts of light illumine the east. Even if we can't hear thunder, by clicking on the AM radio we hear the crackling static of lightning, a sort of transdimensional thunder. Somewhere, maybe over the old mining towns of Granite or Greenhorn, nature is birthing fire.

There are few trees between the two rivers at the southern edge of the Columbia Basin. Actually, there are probably more trees now than before the turn of the century, as settlers imported poplars and fruit trees to serve as windbreaks and orchard stock. Overgrazing in some drainages has encouraged the opportunistic juniper to get a foothold where native bunchgrasses once dominated. Still, trees are scarce, and lightning must be fairly dry to start fires; if thunderstorms aren't dry, the least amount of precipitation will extinguish a newborn fire in light grass. If the fuel is dry and the temperatures are high and the wind is brisk, these fires quickly gain momentum amidst the brushy box canyons and draws beneath the wheat.

There is a certain unobtrusive intensity about these blazes. The only time they really burn well is when dryness thoroughly penetrates the landscape, and under these conditions they can consume hundreds of acres of bunchgrass and sage within what seems like minutes. Typically, the smoke surges up in a broadly deformed plume of gray and white—black, when a heavy concentration of sage or a juniper torches off. If wind driven, the plume bends over and dissipates within a few hundred feet of the flaming front, leaving a smoking, rapidly cooling landscape in its wake. These fires are mature, even if their adulthood looks different than that of other fires. There simply is not enough heavy fuel to sustain the amount of heat and combustion needed to bring on the spectacular maturity of the forest fires that create convection columns tens of thousands of feet in elevation, capping out in milky gray cumulus, tossing firebrands miles ahead of their advancing fronts or forced landward by powerful downdrafts, capable of overpowering local weather patterns in an outdoor furnace pressed in by nothing but vegetation and air. Maturity takes many forms.

These fires along the rivers may scorch several thousand acres and drag on uncontained for a day or two or three, but rarely will they evoke the media interest triggered by the more defined smoke columns that billow up dense and black from Western forests. Where there is little apparent commodity damage, except when these river fires claw their way up out of the canyons and into unharvested grain fields, there is little media concern. Of course, those who winter cattle on the canyon slopes would not share that unconcern, even if *lack* of fire is the very reason that decadent, sequoia-size sage has become a dominant feature of the landscape.

The names of these obscure eruptions won't be found in tomes on fire history or as case studies in national fire training. Nevertheless, these fires have burnt just as deeply into the memory of those whose backyards they occupied and whose lungs sucked in their residue as any of those fires receiving national news coverage. Sixteen Mile, Northpole Ridge, Oakbrook, Two Springs, Hogwild, Chicken Springs, Smith Canyon, Narrows, and a host of other fires burned through the sage and wheat and fescue and rabbitbrush that frame this austere strip of aridity. The tactical events and decisions made on the majority of these and similar incidents tend to coalesce over time. The little Smokey Bear calendar books that I've accumulated over the seasons inscribe the barely legible names of a good many of these fires. Alongside the names are scrawled overtime hours or hazard pay received, but few details of how any given fire was fought or who did what, speculations about why it was done, or the feel and smell of smoldering rangeland. Sometimes I wish I'd kept better records; sometimes I'm glad I didn't.

When some of these fires did make the local and regional newspapers, it was usually due more to local citizens raising a clamor with federal suppression policy than anything inherent in the fires themselves, even if a fire's behavior was anything but lackluster.

The lore and myth surrounding a few of these blazes have moved them into the realm of sacred history—a time in which the extraordinary was a bit more common, where our deeds were a bit more honorable, and where the incompetence of management and our own clearheadedness was a little more transparent. This is also a time that has left little tangible record other than chicken scratches in calendar books and the accumulated traditions and remembrances of those who met the advancing flame fronts. Depending on how you look at it, or how you remember it, the past might be either a golden age or the dark ages. It's debatable whether some of the remembered events actually occurred; whether or not the cluster/fucks of mass disorganization, screw-ups, successes, run-ins with unsavory characters, or moments of glory occurred exactly as remembered is subsumed under the meanings bound up with those events. These remembrances have shaped the collective identity of those whose camaraderie bordered the flames. This is no excuse for irresponsible yarn spinning that stretches the importance of what we and others may or may not have done. But with that as a caveat, these stories still serve as archetypes, both for who we are and who we've become; they are sagas and epics that shape the moral import of what we do, and who we'd *like* to become. They reveal the propensities of our psyches, even if unsubstantiated by hard fact. And like the dichotomy between the Kent farmer's understanding of community and the dilapidated face of the present throughout Sherman County, this, too, is not unimportant.

◆ ◆ ◆

There can be a great deal of intimacy in meeting the margin of a fire that is eating its way through golden-blonde cheatgrass and sage. Unlike project fires that require overhead management teams and sometimes hundreds, even thousands, of firefighters and days or weeks of mop-up, a fire burning moderately through grass frequently lets you cozy up close. Even if it is sprawled over several thousand acres, you can herd it along with a dozen or so people. This, however, may breed a false sense of confidence in the skill of those engaging such a fire; it can also engender camaraderie; it might breed both.

Little known and even less used in fire suppression these days is the burlap bag, or gunnysack. Not so between the rivers. During the summer months, many a farm rig will have a few wet gunnysacks stuffed in a five-gallon bucket in the truck's bed. They jostle around next to a weed sprayer filled with water for summertime fires, a couple of old irrigation shovels with rusted heads and sun-bleached handles, and, possibly, a galvanized backpack pump purchased down at the co-op. We, too, always carried gunnies. I'm not sure who began the tradition for our fire agency; maybe it was someone who ranched or farmed or was a historian familiar with attacking fires on the Great Plains. It certainly has proven useful.

There are few sights more enticing than a fire moving slowly through fine, grassy fuels: gunnysack country. If accessible by road or in terrain that isn't too rough, the fire can be attacked by squirting water from a moving vehicle; but along the lower Deschutes and John Day Rivers, grain fields nuzzle the canyon rims. From there you walk, and when you walk, you carry a tool and a gunny.

Whether sizing up a going fire or just driving through the country, you begin to develop a "what-if" mind-set. "What if a fire were burning on that slope?" "How slow would the fire have to be moving to attack it with hand tools? Engines? Burlap?" In fact, certain terrain and vegetation invite such speculation, and, a fortiori, invite the hope of their actualization. Basic firefighter training drills the rookie with the concept that firelines must be scraped to mineral soil. But on the Columbia Plateau a fireline may also consist of smoldering grass stems beneath a flailing bag of wet burlap.

As you drop down the twisting path of U.S. Highway 126 into Maupin, where our old guard station was located, there is a stark view of Bakeoven Creek Canyon as it carves a path from east to

west, leaking its feeble trickle into the Deschutes River at the
Maupin City Park. There is a certain sensuous quality to the little
ridges between spur canyons that dissect the slopes on either side
of the creek. Some of these clefts are filled with alder and willow
and Magnar wild rye. But it's the rounded hills between them that
are the most interesting. In the evening or early-morning light of
midsummer, the contours are well defined, and the brushy draws
hem in the south-facing hummocks along Bakeoven Creek. The
golden-tanned slopes have a soft look from a distance, enticing
you to reach out and grab them, or even take a bite out of their
faces—if only you were a giant. A secret reverie. Where some of
the side draws hook around and create a north aspect, a few traces
of red, partly cured cheatgrass and foxtail highlight what is otherwise bone-dry fuel. I'm not sure whether these hills are so attrac-
tive because of their fleshy analogues or because of an intrinsic
beauty wrapped up in their gentle curves and arcs. Maybe it's
both. Either way, they are hills that, on fire (or even before they
rustle with flames), cry out for a gunnysack.

Just as the dimension through which you approach a fire weds
you to the fire environment and can be more or less mediated,
the way in which you engage fire in its native home can be more
or less intimate, and there is a time and a place for different
forms of greeting. Slapping a wet burlap bag at the flaming edge
of a fire is about as close to fire's terrestrial habitat as you can get.
Scraping a control line next to a fire's perimeter allows a similar
degree of closeness, as you nuzzle up to its flanks with nothing
but a shovel or pulaski; however, to straddle the edge of green and
black while whipping a gunny—one foot in the still-smoldering
wads of grass roots and ground litter (a fuel-eviscerated safety
zone in which to take refuge if the fire gains vigor), the other foot
in yellowed grass—is an even more intimate and personal way to
engage fire.

To meet the fire, to see the hot flames go black beneath your flailing sack, brings with it a definite sense of accomplishment. I'm sure the same could be said for a constructed hand line, as you witness the advancing fire meet it and go black or your burnout move away from the fire line toward the margin of the main fire, choking its source of fuel in the process. Maybe the same could even be said for a bucket drop from a Bell Jet Ranger helicopter as it trails seventy or eighty gallons of water along a fire's flank, cooling and smothering it to a muddy death, or a c-130 as it belches slimy retardant. Still, there is an unmediated closeness in meeting a fire, on the ground, with a damp chunk of fabric. It's a lot less obtrusive than a D8 Cat gouging a trail down a ridge. And in this country, a dozer scar may adhere to a slope for decades—much more damaging than the fires that have rejuvenated this landscape for millennia. It's a question of whether to meet fire, and the land on which it burns, respectfully—tenderly—or to lord over it, contemptuously. Cold. Jaded. So it seems that the tools we use reflect and affect our souls just as much as they do the land.

There is pleasure in completing little tasks—sweating your way up a hill to the flank of a fire under the sun and open sky of mid-July, then, in the company of several others, swatting out flames until either you smother all movement, or cool, moist night air tucks the fire in for the evening. This genuine satisfaction does not abdicate you from the responsibility of asking why you are doing what you are doing, and why it is meaningful. And of all the seasons of the year, summer, the summer of fire, is when these questions are cured.

5

Engaging

No doubt it has often been stated
that the conquest of fire definitely separated
man from animal, but perhaps it has not been
noticed that the mind in its primitive state,
together with its poetry and knowledge,
had been developed in meditation before fire.

—GASTON BACHELARD,
The Psychoanalysis of Fire

In a given life, moral reflections on fire emerge slowly. As a child and adolescent, one may experience multiple seasons of fire, but the latent questions of why humans find fire so entrancing and its control so alluring often require adulthood and the radiant fires of summer to break open their husks. Once such questions are asked, reaching an ignition point of consciousness, they can be either nurtured and engaged or repressed, rationalized, and allowed to blink out like a neglected hearth ember. My intent in this chapter is to stare ever more intently into the flames of fire and try to articulate the meaning they have for all our lives. The answers I suggest, though lodged in summer, will press forward and loop back to other seasons of the year.

Naturally, there is a panoply of instrumental *uses* for which fire can be employed as a tool, such as cooking, lighting, heating, and hunting, or landscape and habitat modification. But even these uses are shot through with a nonutilitarian residue, an irreducible attraction to fire itself, which I adumbrated in the robust glow of the campfires of spring and the flaming margins of the canyon lands of summer. To explain this attraction, some individuals have taken a backward glance through history to find the patterns, practices, and myths analogous to, and the basis for, modern attraction to flames; others carry this etiology one step further by lodging traditions, myths, and metaphors—ancient and contemporary—deep within the human psyche. While both approaches have merit, I believe that watching, tending, and engaging fire constitute experiences from which personal meaning is derived, and this apart from simple historic or psychological reductions. There is value in what is seen and felt in fire, and this is *good in itself*. Still, the point can be pressed, why are *these* experiences meaningful? To get at this, let me provide a counterpoint in psychoanalysis.

Among those of the psychoanalytic persuasion, sexuality is a popular and often deservedly powerful touchstone. And when we come to the subject of fire, there are, irrefutably, many examples of fire origin myths that credit human genitalia as the instruments of both the genesis and the sheltering of fire. For instance, J. G. Frazer, in *Myths of the Origin of Fire*, relays an Australian tribal story:

> The men had no fire and did not know how to make it, but the women did. While the men were away hunting in the bush, the women cooked their food and ate it themselves. Just as they were finishing their meal, they saw the men returning way in the distance. As they did not wish the men to know about the fire, they hastily gathered up the ashes,

which were still alight, and thrust them up their vulvas, so that the men should not see them. When the men came close up they said: "Where is the fire?" but the women replied: "There is no fire."[1]

Though it is not altogether transparent what role stories such as this one played in primitive societies, is there not reason to think that such myths may have been taken both literally *and* metaphorically by their tellers?[2]

Gaston Bachelard, in his book *The Psychoanalysis of Fire*, undertakes the ambitious project of ferreting out those aspects of human experience, entwined with the unconscious mind, that encumber the search for objective knowledge relative to fire.[3] At one point in his discussion, Bachelard conjectures that early humans drew a comparison between the rubbing of two sticks and the rubbing of genitalia, where the former eventuates in the heat of ignition (and, I might add thankfully, the latter does not). The propensity for humans to link sexualized heat and fire is what Bachelard calls the Novalis Complex.[4] There is some cogency to this. In a world possessing limited knowledge of human physiology, it would have made perfect sense to infer that the burning in one's loins was a literal instantiation of fire, kindled in the genitals. At the very least, the heat associated with both orgasm (or intercourse) and the fire ring might share some more basic constituent binding the two felt realities together. The ancient one might not have articulated it so, but that some such prescientific inference may have occurred seems reasonable to assume.

However, that today's attraction to fire must be reduced to a shadowy substratum of sexual urges (or libido) is less convincing, even if fire as a metaphor for passion is still alive and well in modern culture. I think that Johan Goudsblom is quite correct when he writes of the potential excesses of psychoanalyzing human responses to fire:

Sigmund Freud was certainly right when he pointed out that the acquisition of fire demanded the renunciation of certain spontaneous urges. The only urge to which Freud paid any attention, however, was the supposedly irresistible need felt by primal man when he came in contact with fire "to put it out with a stream of urine." This infantile desire, connected with the enjoyment of sexual potency in a homosexual competition, had to be overcome.

Of course, for any human group to preserve a fire it is necessary that its male members abstain from peeing it out. But our early ancestors surely had other and more urgent problems to cope with.[5]

Indeed! I have no doubt that many a campfire has been peed on as a novel way to extinguish the blaze, but to reduce such an act always to deep sexual motives (though at times this may be true enough) may speak more to the copulative predilections of the theorist than the cogency of the theory. I don't want to discount the very real connections that some contemporary fire infatuation, or pyromania, *may* have with sexual urges. And as the history of human deviance can attest, just about anything may elicit physical and sexual arousal, in the eyes of some persons.

Nevertheless, it is questionable whether a psychoanalysis (of the reductive sort) is needed to acknowledge that there can be layers of motivation and meaning lying beneath ordinary conscious attitudes toward some event or phenomenon. Even Bachelard intimates that there is more to reverie before flames than is reducible to sexualized fire complexes, however strong those may be. The deeper problem for psychoanalysis is whether, in the case of fire, *attraction* can be reduced—in toto—to sexuality. This I doubt.

If for no other reason, such a myopic approach seems dubious in light of prepubescent attraction to flames. Long before they have felt the heat of passion, children find fires alluring. So whatever else one might conclude in constructing etiologies for attraction to fire, unless one goes to the extreme of reducing such feelings to latent sexual energies (which in due time will erupt of their own accord), such childhood infatuation and reverie need to be explained in other than purely sexual terms.

This suggests another and, I think, truer understanding of the motivation behind attraction to flickering flames, a motivation that anyone who has ever viewed the driving front of a wildfire knows: seeing something new. Seeing new things—things often hidden—satisfies our need to know about the world. Just as the glowing coals of a campfire are the medium through which one witnesses an oft-shrouded side of reality, and just as that act of watching transcends the merely psychological in that the seeing latches onto objective qualities of a phenomenon outside (though encompassing) the observer, so, too, seeing something new in the grand scale of wildfire touches both the soul and something beyond, and that "beyond" is awesomely beautiful. Such seeing is a way of learning, and to learn about one's place in the world is surely meaningful in and of itself. The echo of Aristotle is heard in the crackling of freshly lit tinder, for potentialities are being actualized—potentialities whose latent power bursts forth in ever new patterns and intensities.

A child poking a willow switch in a campfire—after the marshmallow has been either eaten or incinerated into a smoking, lumpy glob—soon learns that green springy branches may sizzle, blacken, and flame a bit but generally do not sustain combustion very well. On the contrary, grease-soaked paper plates or split pieces of creamy yellow tamarack will flare up quickly, and cambium that was once alive is now on fire. Such are the paradoxes of fire and life. Green grass smolders; dry grass burns quickly; and even on cold nights, it hurts to get too close to the brightly colored, glowing ring, and standing too far away makes one shiver.[6]

While the fire of the hearth has taken on connotations of safety, home, family, and repose due to the life-sustaining and enrich-
ing warmth that it provides, it is rare to hear of the wholesome-
ness of knowledge gained through encountering the flames of
free-burning wildfires, the fires of summer.[7] Could it be that felici-
tous feelings toward destructive phenomena (which is to say,
prima facie evil) ought to be suspect? Probably. And in the case of
wildfire, this seems especially justifiable, for wildfire does consume life, and, at the very least, life embraces the moral realm.
First impressions, though, don't always bear up under scrutiny,
which I believe to be the case relative to wildfire.

To play the devil's advocate for a moment, though, even if we
allow that the knowledge one gains in encountering a wildland fire
may be wholesome, finding attraction and meaning in such experiential knowledge *can* be warped. This is exactly the fear of those
who recoil from affirming the aesthetics of fire. Imagine an aesthetician-scoundrel, a person entranced merely by what is aesthetically pleasing, fixated on ever new arrangements of color, light,
and shape, and these qualities cut loose from other, moral considerations. There is always yet another, even infinite, variation on
visual themes that could be imprinted on one's retinas. For instance, after such a deviant one has viewed the glistening crimson
hues of a human torso blasted open by cannon fire, there are still
many other color arrangements to investigate, like the contrasting
hues of freshly exposed bone and bile, or what such images might
be like if vivified by flames. There is also the unfinished business
of discovering how different methods of disemboweling affect the
"beauty" of the scene. This, however, is only partial seeing. It is
knowing based on only one facet of value.[8]

There is a diminishing return on the value of *some* knowledge
gained. In such cases, the content of the knowledge gained ought
not to be known. There are other more important values super-
vening on lesser ones, such as experiencing the satisfaction of re-

specting the well-being of a human being over and against titillat-
ing one's eyes and satisfying one's artistic curiosity. Though less
shocking than the hypothetical aesthetician-scoundrel, could it
be that there is a link between seeing something new and the
magnetism of horrifying or gruesome movies? During such spec-
tacles, one witnesses what one (generally) hopes will never be
personally experienced, though one may secretly wonder what it
would be like.

As distant as this may at first seem from the entrancing power of
fire, there is an analogy. I contend that depraved and wholesome
inquisitiveness both draw from the same province of the soul. This
somewhat parallels another psychoanalytic complex identified by
Bachelard—the Prometheus Complex. For Bachelard, this complex consists of "all those tendencies which impel us *to know* as
much as our fathers, more than our fathers, as much as our teachers, more than our teachers."[9] I know very well the attraction of
seeing something new and big and momentous in a wildfire, as is
attested by my extensive library of slide transparencies and photographs of fire. This, I think, merely reflects the truth that a multi-
tude of images burn into one's memory in the face of smoke and
firebrands and blazing tree crowns, and that in order to revive and
share memories of such images, they must first be burnt into a roll
of Kodachrome. Such is the case whether it be a dense, black,
August smoke column, the contortions of a fire whirl bending and
bowing in the eclipselike darkness of a soot-shrouded midday sun,
or the yellow- and green-clad individuals, eerily out of place as they
stand silhouetted against waves of orange and vermilion, with tools
in hand, like so many migrant workers going out to the fields.

Since fire is a happening, each new blaze is utterly unique
and will never be replicated, spatially or temporally. The fire I see
today is not the fire that I'll see tomorrow or saw yesterday. Something irreversible is happening, including my experience of the
happening. As Johan Goudsblom writes: "It is impossible for the

remains [of fire] to revert to their original shapes and colours. The phoenix rising from its ashes exists only in the human imagination."[10]

And the more uncommon the sight, the more the desire to capture it for posterity—if not on film, then at least in the mind's eye. By staring long and hard at a fiery spectacle, one satisfies a deeply ingrained curiosity; unfettered by moral sensibilities, such gazing becomes mere lust, a lurid voyeurism. I suppose that such a fixation is not limited to fire that is wild, for one could relish the act of stuffing sundry objects into a fireplace to see their varied modes of combustion. But wildfire can be spectacular. People flocking to see the power of fire is no different than people congregating to view the height and speed of a swollen river at flood stage—half hoping that old flood records are being broken.

During the 1994 blowup of the South Canyon Fire, one firefighter asked another to snap a picture of the enlarging flame front because "that was his brother-in-law down there and his family would never believe this."[11] The two firefighters on the ridgeline barely escaped to develop the film; those captured in the frame, including the brother-in-law, did not. Yet the irony is thick, for at South Canyon, and just about anywhere else that wildlands erupt into fire in the receding sun of the summer solstice, there is beauty in the flames. Surely, trying to capture a fellow firefighter against the pumpkin-orange glow of thick smoke and flame may be a way to brag about one's manliness. It may also stem from the desire to immortalize the experience of being present to hidden powers in nature becoming explicit.

Even individuals still nursed by the sentimentalized nectar of the Ad Council campaign that features Smokey Bear occasionally admit to fire's aesthetic dimension, though the unease with such feelings often remains: Are these feelings moral? This question is especially sharp when juxtaposed with the fact that a few champions of flames' aesthetic qualities express their infatuation with fire by sneaking around torching tracts of Southern California chaparral. Must the appreciation of beauty in flames, and the experiential knowledge gained from such appreciation, always be linked to the deviant (sexually fixated) arsonist? Again, I think not.

To see a violent conflagration firsthand is to bear witness to what is relatively uncommon on a day-to-day basis and altogether absent during some seasons of the year. Though related to the same basic urges that tempt one to take a second glance at an automobile wreck, or to go out of one's way to view a house fire, neither of these latter events are good in themselves. We would all be better off if no one got in car wrecks, even if this left us bereft of knowing what mangled steel and shattered glass looked like when two cars collide at high speed. Similarly, we would be no worse off if houses never burnt down, even if arcing electrical lines, exploding propane tanks, or the flashover in a superheated room can make for spectacular images. Both house fires and car wrecks always have a tragic human dimension entailing little in the way of redeeming value. Wildfire can exist and has existed apart from the human touch and, even when unleashed by humans, has continued to play a vital role in the ebb and flow of the natural world. Must black always be ugly and green always be beautiful?

But what of the experiential knowledge gained from more active forms of fire engagement? Certainly, some of the more ethically dubious examples of taking pleasure in seeing something new occur when human participants actually bring about that newness. It's in the delight one feels in setting ablaze hundreds of acres of rangeland or forest—during a backfiring operation to remove fuel between the mineral-soil control line and the main fire, or in a prescribed burn—that the moral ambivalence of seeing beauty in fire again rears its head, and the tacit fear of becoming a professionally sanctioned arsonist wells up.[12] To say that this isn't a moral dilemma is both callous and flat wrong, for the living of a truly good and meaningful life as a human agent involves being able to exert some degree of control over one's surroundings. However, as with the aesthetician-scoundrel, not every exertion of power for the sake of knowing is ultimately good, even if the ability to wield power remains a genuine nonmoral good of sorts. It is no different with fire. A great sense of power accompanies sending forth a mile-long fire front from a thumbnail-size flame. With fire in hand—be it a torch mixed with gasoline and diesel or the searing blast of a fusee (the firefighter's version of a road flare)—we can markedly extend our grasp over nature. This presents a striking contrast to our day-to-day comings and goings in which we are usually able to influence only a small sphere around us, verbally or physically. With mechanical help, we can exert more control over our surroundings. With fire, we break past appearances, beckoning forth energies that continue to expand and roam regardless of our wishes. The genie cannot be gathered up as easily as it is released, and in both the setting and the consequent watching, we find meaning.

Even granting that some fires are best controlled through lighting backfires, or that the best way to return nutrients to some soils or reset a chain of succession is through firing the landscape, one may still be uneasy about feeling satisfied in kindling such blazes. Is one taking pleasure in destruction *for the sake of destruction,* or is one pleased knowing that what is consumed in the flames is destroyed for positive ecological effects? Is the latter a mere justification for the former? Is the state of mind in such acts linked to an ontic and normative substratum of value? I'm sure the end result of satisfaction in watching and kindling is some blend of the two motivations. Further, I want to suggest that a more charitable appraisal can be given to engagement than may commonly be the case, and this apart from cheap rationalizations.

Without buying into the folly of moral relativism (as if good or bad have meaning only in the eyes of the beholder), one can still claim that in many actions it's not so much the *act* itself that is wrong but the *reasons* for doing it. Take, for instance, the case of those carnival-like fund-raising extravaganzas that charge you five dollars for the chance to swing a baseball bat at a car. Sure, the car may be nonfunctioning and even destined for the wrecking yard; sure, the money is going to a good cause; sure, swinging a bat into the windshield of a lemon may be better than letting off steam by smashing in a person's cranium. But the delight taken in destroying something just for the sake of destroying it, or just to see how much force it takes to cause such and such an amount of damage, seems morally suspect. If this is where meaning and fullness in life reside, then the real evil in the world would be lodged in not having a constant supply of things to destroy. Similarly, any setting of fire over which one merely gloats about the amount of destruction accomplished—as if ashes and charcoal are prized trophies—is also suspect.

Maybe there is the need to do something that seems to matter, something that leaves its mark on the world and gives one the sense that one is not insignificant. This is a legitimate need, without which life would seem like a protracted joke of little or no significance. There can be a genuine sense of pleasure and power in jamming the dogs of a chain saw into the scalelike bark of an old yellow-bellied ponderosa, pivoting the whirling, bladed chain deep into the tree's core as curled, light-yellow chips pile up around the tree's butt. Something significant is happening: A person only a few decades old is bringing to the ground a mass of bark and wood and needle that clawed its way into the sky over the course of a hundred years or more. In a few minutes, the tree's patient ascent is severed, and once the ponderosa drops to the ground, its life is finished. Taken in itself, such an act may be the right thing to do, but it could be a darkly immoral act if the adrenaline rush accompanying the feeling of power is the prime mover. The same logic applies to unleashing fire, and to reveling in the new sights that the enveloping flames provide.

Finally, maybe wielding power over nature stems from an urge to press the boundaries of the forbidden and to play with the taboo, regardless of what the prohibition may be. To do what one wants, and to do this in defiance of another, may seem doubly powerful, especially with a phenomenon such as fire. Take the example of backfiring, again. That setting a backfire is permitted within the scope of the firefighter's job, and within the strategy and tactics of suppressing a given fire, does not sweep aside moral ambiguities any more than a sociopath who longs to kill would be without blame for killing people during wartime, even if such killing were legally permitted by some government.

Fire, like any other human-initiated event, requires respect for the thing it is and the role it plays or could play. To unleash fire for a narrow range of merely selfish reasons—to promote pretty colors, to indiscriminately exert control or power over some little chunk of the world, to destroy life for the sake of destroying life, or to fondle the taboo—is morally problematic. The uneasy twinge felt in the gut by anyone who has lit off some part of the landscape attests to this moral ambiguity. But moral ambiguity does not equal immorality. Seeing new things *can* be good, as is freely initiating change in one's world. These are legitimate needs, and whether or not we fully comprehend the psychic underpinnings of these needs, that they *can* contribute to wholeness and vitality in human life seems reasonable and experientially justified.

To see new things and to embark on a journey in life in which one attempts to engage in pursuits that really matter reflect a heroic stance that, to varying degrees, is part of all human experience. (In chapter 7 I'll evaluate more fully how this heroic journey manifests itself in the face of flames.) To engage—even to dance

—with fire encompasses more than simply fighting it for the sake of extinguishing its heat and light; engagement is concerned with confronting fire in its moods and modalities, which are always new and always significant.[13]

Entering deeply into the natural cycle of life by igniting or suppressing fires can be a good thing, good in the end result, and good in the meaningful control it provides human agents. Thoughtlessly suppressing and withholding fire from nature, or indiscriminately setting nature ablaze, can shut down life-promoting cycles and drastically change fire regimes; unleashing fire —or allowing it to be unleashed by nature—can also engender a profusion of life in a beautiful display of shimmering energy. Goodness or badness of motivations and outcomes in fire engagement encompasses all these considerations, and none is completely reducible to components of the unconscious mind.

◆ ◆ ◆

The sun's heat is fully matured now. As the blinding white disc arcs across the sky in this season of sun and fire, it serves as a gauge both for the lives of those who engage fire and for the lives of the fires that have the potential to mature from infancy to adulthood in a single day. Summer is the time for such things.

The environment that fire inhabits is itself a complex source of knowledge for those who deeply engage fire. Take weather. The towering cumuli are powerful sights to behold; the surging electrical currents that connect cloud to cloud and cloud to ground are dramatic testimonies of nature's ability to evolve something new. It doesn't take long for a firefighter to learn that, when the clouds begin building by early afternoon, there is a good chance that by the time the sun dips beneath the western horizon, little "suns" will spring forth out of broken-topped pine snags and hoary old junipers. And as the great solar disc goes out of sight, leaving its convective remnants billowing in cells of cloud and flashes of light, small terrestrial lights grow ever more luminous in their own march toward adulthood.

Through great effort, first in lightning and then in terrestrial fire, day presses into night past the closed arc of the sun. Taken in its entirety, this is an object of true seeing and meaningful knowing for both the watchers and engagers of summertime fire. Recognizing the sinews that bind and draw this extension of day forward into evening, and then night, is what enables one to press beyond the lust of the eye toward the appreciation of wholeness. Yet in the wholeness, there is tragedy.

6

New Shoots

Dr. Hawkings, the physician who went in with
the rescue crew the night the men were burned,
told me that, after the bodies had fallen, most of
them had risen again, taken a few steps, and
fallen again, this final time like pilgrims in
prayer, facing the top of the hill, which on that
slope is nearly east. Ranger Jansson, in charge
of the rescue crew, independently made the
same observation. The evidence, then, is that at
the very end beyond thought and beyond fear
and beyond even self-compassion and divine
bewilderment there remains some firm
intention to continue doing forever and ever
what we last hoped to do on earth. By this final
act they had come about as close as body and
spirit can to establishing a unity of themselves
with earth, fire, and perhaps sky.

—NORMAN MACLEAN, *Young Men and Fire*

According to the calendar, the first day of spring occurs several weeks before Good Friday, a symbol of new life leading up to penultimate death. For a few who mark New Year's Day with cloud and lightning, their first day of spring will begin a sinuous Via Dolorosa, not to Golgotha but to their own consumption under the pall of rolling smoke and all-purifying flame, and this for the sake of snuffing out the last vestiges of summer. This path has eventuated in names such as Setzer Creek, Blackwater, Mann Gulch, Rattlesnake, Inaja, Loop, and South Canyon. No multiplication of fire orders or watch-out situations has or (by extension) will completely eliminate pilgrims on this path, for fires will continue to burn, and humanity will continue to engage fires, and people sometimes get in the way.

Each generation is jolted by some such pilgrimage. The degree to which the tremors are felt varies from person to person: Some feel as if they are rocked by the epicenter, while others look on at the radiating margins. Wherever one is located, though, the locus of shock remains the same. For me, the path of finality that resonates most clearly leads southeast, to Colorado.

◆　◆　◆

In the shade of the Gambel oak, they died. Whether or not the tangle of fire-cured, golden-brown oak brush provided natural shade from the July sun was of no consequence, for the fire that erupted at the bottom of the ravine, hopscotching up the slope through both green and half-burnt oak, juniper, and piñon pine, provided shade of its own. Amidst the blinding light of flames hundreds of feet high, fourteen people perished. Though always a possibility for a wildland firefighter, such an outcome is never viewed as an inherent necessity; in theory, and unlike war, battle can be waged along the fireline without casualties. Nevertheless, on 6 July 1994, firefighters moved up a slope to their deaths in the shade of smoke and flame on an obscure ridge of Storm King Mountain in the central Colorado Rockies.

Six weeks after the incident on Storm King, I traveled to Colorado, partly out of reverence, partly out of curiosity. The furnishings of fire's native home are multivalent, and having engaged the margins of fire in a variety of terrain and vegetation types, I wanted to see with my own eyes what mix of these furnishings could actually lead to the death of firefighters—firefighters who also called Prineville home. Sure, a simple campfire generates enough heat to kill, but rarely (except in the case of an intoxicated stumble over a rock ring) do campfires kill. I suppose any wildfire *could* kill, so in reality I need look no farther than my back door to have this curiosity sated. In August 1994, the hypothetical interested me little.

In search of a topographic map of the Storm King Mountain/ Glenwood Springs area, I stopped at the BLM district office in Grand Junction, Colorado. I asked the woman at the front desk if they had any Glenwood Springs quads.

"Are you going up to the mountain?" she queried, with a cracking waver in her voice.

Maybe she knew why people requested Glenwood Springs quadrangles these days, or maybe she saw the patch on my sweat-stained, royal-blue baseball cap that read Prineville BLM Fire Management. Either way, she was right, and I probably wasn't the first person from the Pacific Northwest who had wandered into her office in the last month.

"Yeah," I said.

I asked how much I owed for the map. "Nothing," she answered.

I thanked her, walked out of the office to my car, and continued east, thinking about *the mountain*—a little-known indefinite article of the landscape now become definite.

It was a bright sunny drive from Grand Junction to Glenwood Springs on Interstate 70—past multicolored clumps of sandstone, through the little towns of Rifle, Silt, and New Castle. And as I came around a bend in the freeway, there it was—the slicked-off slopes of Storm King Mountain that I recognized from photographs and TV footage as the site of the South Canyon Fire.

I exited the four-lane ribbon of concrete that paralleled the Colorado River and parked my car as close as I could to the base of the ridge that I figured was my destination. The afternoon heat felt heavy as I emerged from my air-conditioned car and strapped on my IA (initial attack) pack, which had accompanied me for the last several seasons of firefighting. It was light, containing only maps, a copy of the "South Canyon Fire Investigation," water bottle, camera, and film. In preparation for the hike I stripped it of fusees, fire shelter, an MRE (Meal Ready to Eat), and a fire-resistant brush jacket; I wouldn't be encountering any flames on this day. In retrospect, though, I wonder if I cheated. Why should I carry less weight than those who carried full packs but would never carry packs again?

There were a few strips of fluorescent ribbon hanging from a bush near my car. I followed them, and like the bread crumbs marking the way home in children's stories, more orange ribbon fluttered from oak and piñon branches, angling to the northeast along a freshly beaten path.

The ground was very dry. There was little surface fuel where the trail meandered, but what brush, downed limb wood, and yellowed grass was there was dry—a punky, crumbling, rotten dryness. Even the standing juniper and piñon looked thirsty under the August sun. I tried to imagine carrying a tool or chain saw up the slope—my own physical burden was so light, yet my throat was parched.

I hiked up the ridge to the west of where the fire was located before the blowup. Eventually, I broke out of the green piñon, juniper, and oak, and into fresh blackness—the western flank of the fire, where it was halted. Now turning to the east, I descended through powdery gray ash into a ravine and then climbed the opposing slope. An hour and a half after leaving my car, I arrived at the fatality site.

On the way, I ran into the original line that the firefighters had carved along the slope before the blowup. Their control line was not too hard to find: like rows of pungy sticks, charred oak stobs lined a foot-wide cup trench, built to catch rolling embers from crossing the line and starting spot fires, that is, new spots of fire outside the control line. The stobs were sawed at angles and left sticking up several inches above ground. Someone made this; someone with thought and skill directed a saw along this angle; someone pulled a pulaski and scraped a shovel along this little trench. Life seeped out of several humans here. Hallowed ground.

Among six- and seven-foot-tall blackened oak skeletons, there was an amazing display of fresh plant life. It had been only six weeks since the fire, but already thousands of small, light-green Gambel oak shoots had groped their way up through the talcum-like ash. Under the hot August sun, spring crept out of the hillside; spring always follows fire.

Close by, and about the same height as the infant oaks, were a dozen heavy-gauge spikes driven into the ground. They were strung out up the slope to my east some hundred feet and had small, round brass tags wired to them reading FF #1, FF #2, FF #3, and so on. "FF" could mean many different things in other contexts; here it clearly meant "firefighter."

I wasn't alone. There was a small group of people milling around on the ridgeline above me. A few were clad in yellow Nomex fire shirts, green fire pants, and variously colored hardhats. Most were dressed in everyday clothes. Later, when I made it up to their location, one of the people wearing Nomex told me that these folks were relatives of those killed. It seemed good and right to fly the families up there.

With the exception of steel spikes, small brass tags, and several bouquets of sun-withered flowers, nothing much distinguished the aftermath of this medium-sized wildland fire from that of any other fire that had burned on the west slope of the North American Rockies since cured vegetation and lightning were first betrothed countless millennia ago. Life of one form or another always perished: first, the plant that was struck by the lightning bolt, be it tree, shrub, or grass; next, fine fuels such as grasses, forbs, needle litter, and twigs or duff; then, if enough heat was generated, heavier brush, still more grass, and even live green vegetation; finally, animal life unable to escape the superheated gasses, smoke, and flame. Little is changed today. When the fire raged where I stood, the same age-old constituents of fire mixed to sustain free-burning combustion—heat, oxygen, and fuel—helped along by the interplay of topography and weather.

❖ ❖ ❖

Since that clear, hot day in August when I stood on the slopes of Storm King Mountain, the oak shoots have surpassed the temporary steel spikes in height. Nevertheless, the presence of both, rooted in the ash next to each other, hints, however dimly, at possible moral dimensions of fire—moral not just in terms of whether or not it is right and good to fight fire, but also in terms of fire's peculiarly formative role as both a framer of meaning and an icon of ambiguity. This is moral epistemology and revelation—an entry point into understanding who we are, what sort of world we live in, and how we come to *know* what is inherently valuable about each.

The fruit born of the ashes speaks of fire's intrinsic goodness or badness, reflecting moral dimensions of a natural phenomenon that can act as the catalyst for growing both verdant oak shoots and rusted steel spikes, for fire promotes vegetation and life, even in the face of death and disintegration. When fire kills some plants, others (resistant to or dependent on fire) gain access to precious sunlight, and a process of succession unfolds. In turning a punky log into a heap of ash, fire unlocks nutrients pent up in the rotting growth rings, giving them freedom to return to the soil from whence they came. By roasting certain varieties of seedpods, fire arouses plant species from their dormant slumber to begin life anew. From the tropics through the temperate zones to the sparsely vegetated Arctic, life in all its profusion has been shaped by fire.

But it's especially in light of fire's consuming human life that we are reminded of the intimate marriage between fire and all terrestrial life. In those instances where life is lost by fire, did the fire kill? In Colorado, was it the drought-stricken Gambel oak, which no one gauged to be nearly so flammable, that was the real culprit? Then again, maybe the dry cold front that swept in from the northwest, fanning embers into a roiling conflagration, should be indicted for murder. And what about the complex web of human decisions, expectations, policies, habits, and self-confidence that only hindsight presumes to have been negligent, overzealous, or inadequate on 6 July 1994? Damn the slope, damn the wind, damn the heat, damn the brush! Who ordered those people to go in there in the first place? Why did they go? Can we judge?

Locating responsibility for tragedy is risky business—definitely in the case of humans, even more if evil resides in a universe governed by deity. In the case of death by fire, human choices are always contributing factors, for it is only by virtue of a whole chain of decisions that one ends up in the proximity of a wildfire—a force of nature that may or may not behave as one hopes. Life is a complex affair, and choices are complex events, yet no less complex than the polygamous relationship between fire and its terrestrial partners—fuel, weather, and topography. To simplify the interplay amid choices made, fire behavior, and tragic outcomes is to do a disservice to both the power of nature and the complexity of human decisions firmly rooted in finitude and fallibility, and not easily collapsed into neat equations.

Nevertheless, in the end, superheated air scorched the lungs of all mammals caught in its searing blast—and there were ashes, and there were steel spikes with small brass tags, and it was the sixth day of July, and it was not very good. By the seventh day of July there were stirrings in the earth that a month and a half later would greet me as miniature oak bushes.

And amid the infernal heat, choking gasses, and blinding flames on the mountain and the irresistible photo opportunity lie clues to the moral ambivalence, and translucence, of wildfire.

7

Man Against Fire

A campaign fire is just like war,
and war, as we all know, is grim.
It's a mighty battle of men and machines.

—DOUG MAXWELL,
fire boss, Canyon Creek Fire

Four miles up Sage Hollow, I turn my four-wheel-drive fire engine off the washboard county highway and continue east on a Bureau of Land Management two-track. Tammy Wynette fades in and out of the cab. Next to the turnoff, a waist-high, brown plastic stake wags in the breeze—a BLM road sign, hardly justified by the grown-over parallel ruts it marks. Before we go too far, I stop. Tim, my engine partner, gets out and locks in the front hubs, insurance for the inevitable mud hole, sand pit, or boulder patch.

June and July came and went with few opportunities to search for fires; now it's mid-August. Two days earlier, lightning boiled out of the south, leaving a trail of blips on our dispatcher's lightning detection screen and small puffs of smoke that kept nearby lookouts busy shooting azimuths and working up legal descriptions. One of the legals is written on a little yellow notepad suction-cupped to the windshield above the dashboard: T19, R17, S4, SW SW.

The road we're on zigzags across Rodman Rim, an east-west juniper-covered scabflat that drops sharply to the north into the Bear Creek drainage of Central Oregon. Even farther to the north lie the forested mountains of the Mauries, Ochocos, and Blues; our BLM lookout atop Tower Point; and the Paulina Fire Guard Station. South of Rodman, the terrain trails off into more lava scabflats, occasional junipers, pronghorn, cows, and acre upon acre of sagebrush. Six miles south of Rodman lies the Millican Valley, part of the Brothers fault zone, and the northwesternmost edge of the Great Basin.

The majority of summertime fire engagement subsists along the edges, such as the geographic edge we presently bump along in search of smoke—transitional moments of various length and breadth, marked by preparation, demobilization, and expectancy; by urban interface, meadow, creek, control line, and the sky blue tinge of Lightning Activity Level 1. Summer is when fire reaches adulthood, setting out on its own, finally able to fend for itself. But, even in a prolific summer of fire, the ordinary has a habit of horning its way into the extraordinary. In a slow year, the edges of downtime stretch out interminably toward threatened layoff dates. Edges are a time of ambiguity. Today, Tim and I inhabit such an edge, on the elongated boundary of a truly ordinary summer, on the edge of the Great Basin, and in the afterglow of a summer lightning storm.

✦ ✦ ✦

For many—maybe even most—Americans, our trek across Rodman Rim is far removed from their conceptions of wildland fire, full of transformation, conflict, struggle, heroism, even war. A just war. Up to a point, these images are accurate. Anytime fire exists, there is transformation—transformation of the fuel as it evaporates in misty curtains of orange and red, and transformation of the microcosm of life and air surrounding the flames. And anytime wildfire confronts human culture, there will be conflict— conflict over how much change fire may invoke, conflict over where fire may exist, and conflict over where fire must retreat.

The history of philosophy and religion is replete with examples of individuals espousing the necessity of trials, challenges, and conflict as means to achieve virtue and meaning—the hero's journey. Some see the field of battle as the necessary topography for this journey. To the extent that human conflict with wildfire seems like war, could fighting fire also be a proving ground for the hero? In other words, is engaging fire really like war, and would we be impoverished without either?

New England is a long way from the cow shit–plastered ribbons of dust called roads that my truck bounces in and out of on Rodman. But there are connections. Some commentators suggest that Harvard pragmatist William James's 1910 essay "The Moral Equivalent of War" both reflects and supplied fodder to a social climate at the turn of the century that readily embraced firefighting as a heroic and virtuous endeavor.[1] James wrote in opposition to those who saw war as a factual necessity, whom he describes as believing that "if war had ever stopped, we should have to reinvent it . . . to redeem life from that of degeneration." These militarists, writes James, "take a highly mystical view of their subject, and regard war as a biological or sociological necessity, uncontrolled by ordinary psychological checks and motives. When the time of development is ripe the war must come."[2]

What James proposes, though, is that even if "our ancestors have bred pugnacity into our bone and marrow, and thousands of years of peace won't breed it out of us," such pugnacity can find a more fitting home pitted against nature in the "moral equivalent of war":

> We must make new energies and hardihoods continue the manliness to which the military mind so faithfully clings. Martial virtues must be the enduring cement; intrepidity, contempt of softness, surrender of private interest, obedience to command, must still remain the rock upon which states are built. . . . If now—and this is my idea—there were, instead of military conscription a conscription of the whole youthful population to form for a certain number of years a part of the army enlisted against *Nature*, the injustice would tend to be evened out, and numerous other goods to the commonwealth would follow.[3]

I'll grant that, if forced to choose, taking up arms against nature is probably a better trail to virtue than shooting or bayoneting other humans. But if the former results in wholesale destruction *for the mere sake of destruction*, then all that is morally equivalent between the two mediums of battle will be their inevitable carnage.

James's moral equivalent, however, seems more nuanced. There is a heroic way to look at engaging nature (or, by extension, fire) that finds meaning in conflict, but only conflict that leads to wholesome consequences. Here, being enlisted against nature might amount to setting out on a quest, a rite of passage concluding in the youthful population's reunion with the communities they left. The initiates will learn lessons on this journey that will be carried home and applied to life's inevitable struggles. Cultures throughout the world have instantiated this hero motif in various forms, and its widespread recurrence undoubtedly reflects internal verities of the human soul.[4] But to what degree engaging fire is praiseworthy, even heroic, is hardly clear. Still, this urge to be initiated, to do that which matters, to struggle, to overcome, and to reunite possesses a degree of moral legitimacy that James's militarists do not.

It's hard, perhaps impossible, to imagine human life gutted of any challenge, possibility for hardship, or external resistance. Indeed, utopian visions that exclude these elements might have more in common with hell than with heaven. But even if such experiences are important to the living of a meaningful life and the nurturing of virtues (which, I think, James would affirm), the next question is, What are fitting contexts to foster such ends? Clearly, war for the sake of war is not among them. Nor is disemboweling nature through clear-cutting, strip-mining, or other consumptive methods *necessarily* virtue producing, especially if what is destroyed possesses value in its own right, worthy of respect. Even though James does not completely spell out what he has in mind with enlisting people "against Nature," there is every reason to believe that kindling, tending, and suppressing fire might be logical extensions of his train of thought. But the burgeoning fire suppression community of the early twentieth century went beyond simple moral equivalence with war: Fire not only constituted a part of nature for conscripts to be enlisted against or engage in, but also was an enemy to be fought.

Wildland fire suppression in the twentieth century was shot through with examples of firefighters, commentators, and fire suppression agencies embracing military jargon to describe engaging the fires of summer. This should be no surprise, given the ubiquity of the term "firefighting." There are similarities between engaging fire and engaging in war: Tacticians set goals and identify boundaries to contain or exclude an unwanted force; personnel sweat, advance, retreat, map safety zones, and save lives; and fire and war call for immediate—oftentimes emergency—responses. From time to time, active military and National Guard personnel are deployed on wildfires. Ex–combat pilots play invaluable roles in wildfire air operations. Still other veterans emerge from their military tours of duty to join firefighting guilds, in which they find a similar crew spirit and camaraderie to that of their platoons, a camaraderie they would be hard-pressed to realize in other civilian professions. And surplus military equipment has always been a fixture on wildfires. Twenty years ago, I would have clawed across Rodman Rim in a drab-green, Korean War–era deuce and a half; today, I'm mounted on a shiny new F-350 Power Stroke.

Yet for all their similarities, war and firefighting are different, which is one reason why I hesitate to call the confrontation with fire in the wildlands a "fight." There remains beauty in cyclonic fire whirls of burning vegetation spiraling hundreds of feet up into the hot summer sky; it's harder to rend beauty out of mangled heaps of blood, bone, and flesh.

The possibility of death by fire probably does reinforce the analogy between fire suppression and war. However, that folks die in the line of duty while engaging wildfire is only minimally analogous to war. Only when fire is cast in the role of enemy does death suggest that the fireline is a battlefield.

I suppose that the term "engage," too, might conjure up hostile overtones. Nevertheless, to engage a thing, unlike fighting it, can connote a healthy intimacy—stimulating, invigorating, full of contact, sensuously rough. To fight requires engagement; to engage does not necessarily entail fighting. Whether one views fire as a happening to fight or to engage may rest on whether fire is seen as an enemy. As an enemy, however, fire is unconscious and forever mute to articulating a declaration of war. More than anything, we're at war with that which inconveniences us.

Even the sense of thrill, or adrenaline rush, often cited as a reason that individuals volunteer for dangerous occupations does not do justice to the personal significance of engaging the wildness of nature found in suppressing or sending forth fire. Such a view settles for a hedonistic answer to the deeper moral values in engagement, reducing complex human motivations to physiology or the unconscious mind. Maybe, though, there really is something valuable about the object of engagement that, while producing a sense of excitement, is not sought simply for some feeling. Similar feelings of nervousness, excitement, and thrill may accompany the paramedic, police officer, structural firefighter, or foot soldier, but to lament the end of sickness, crime, house fires, or war because of their ability to produce "positive" (adrenaline-induced) physical and psychological states is ludicrous, and surely immoral.

Fire is a force, and a happening, that has always been a part of biotic communities, and no less so once our distant ancestors captured the flame. To live in a world that requires our active participation to survive and flourish can, at times, seem a curse; it can also be a great blessing to have a hand in one's continuing to be and become. To utilize nature's gift of fire, or merely to be in the presence of the gift, brings material and even spiritual goods to the human psyche: To see a fire wending its way through a watershed of diseased Douglas fir is confirmation of nature's ability to purge itself; to throw up defenses of shovels, pulaskis, and retardant bombers in the wake of such a fire when it threatens a small town attests to humanity's great endowment of choice and agency. (Where should we live? What is a good way to fashion our lives in relation to the land? What should that land look like?)

Hiking up an obscure canyon aglow with fire, carrying nothing but a backpack and tool, or pecking our way across Rodman without itinerary or schedule is also adventure. Like the mythic hero, those who engage wildland fire temporarily, and for a season, jettison the trappings of society in exchange for exposure to forces that test stamina, intellect, and soul. These sojourners trade familiar communities for a land and time of liminality.[5] Eventually, they reenter the world from which they departed, changed, one hopes —reinvigorated and enlightened. Some do not return. Some return unchanged, which suggests they may not have fully de-

parted in the first place. The journey of the latter is one under-
taken wrongly, viewing enemies as friends and friends as enemies,
a path of ignorance and self-denial.

And while one is on a quest, what really is adventure? In everyday terms, isn't it seeing new things and exploring what is, in one's
own experience, uncharted terrain? Isn't it performing feats that
seem to matter, even if that amounts to keeping oneself alive
against inclement weather, the force of gravity, or a surging brush-
fire? The adventure of confronting and engaging fire is more than
recreation. Engaging fire involves a genuine aspect of creation,
staying in step with a force that is changing the landscape at the
rate of feet per second and chains per hour. In the overall scheme
of things, however, the change wrought by fire is no more significant than the equally miraculous layering of millimeter upon millimeter of growth rings, year after year, in the quiet ascension of a
pine.

✦ ✦ ✦

Our two-track across Rodman weaves through stands of gnarled
old juniper, some with bases up to forty-eight inches or more in
diameter. These are the kind of trees that are most receptive to a
lightning bolt's lurid tickle, and they are also the nastiest to extinguish: interiors rotten and hollow and, once ablaze, filled with
molten punk; trunks disfigured and split; a jungle of branches, all
tugging in opposite directions, from which sawyers do their best
to judge which bearing of gravity will have its way. From the
branches of these old junipers hang dendritic beards of fluores-
cent-green lichen. The ground encircling their trunks is carpeted
by thick, bulbous mats of moss that fill in between pockmarked
chunks of basalt, like grass between garden stepping stones. In
other places, where there are no rocks, the spongy moss radiates
out around the trees' trunks only about as far as the branches
provide shade. And directly under the trees, on this mat of moss
and duff, there is a blanket of rust-colored needles.

Our smoke was turned in two nights ago by Tower Point lookout, seventeen miles to the northeast. Yesterday, an engine crew searched the rim in vain and, at one point, even caught a whiff of smoke before finally giving up. Eventually, the lookout lost sight of the smoke's nibbling rise-form. Was it out or merely lying low?

We drive through squatty, grizzled junipers, clumps of decadent sage, and burnt-out snags in search of a fire that has eluded our predecessors' grasp, and we do so with the hope that a little heat is still entombed in the snag's stump. The smoke we're looking for would take a fairly low priority during a busy fire season; today, however, we're living along that elongated edge of downtime. So we continue searching. We occasionally glass the horizon for haze or puffs that might belie the smoldering snag's readout. We also keep the windows rolled down, just in case a breeze delivers the familiar aroma of burning bark and rat shit.

An excursion like this is a portrait of irony. As the rig's granny gear lifts us up and down over what for all practical purposes are small boulders, and as we grimace at the screech of rock scraping oil pan and differential, we drive past the many casualties of previous lightning storms. This area is peppered with the slumped-over, burnt-out carcasses of junipers. Typically, these trees consist of blackened trunks that rise up six to ten feet, at which point skeletons of outstretched branches droop out from their still hollow cores, a limp, fire-induced impotence. Some snags—by virtue of their jet-black, charcoal-skinned trunks and golden, fire-seared needles—were probably struck down only a year or two ago. Others, mossed over, weathered gray, and speckled with orange and white lichens, have nakedly battled the elements for a decade or more. Most of these burnt snags are solitary, and they are often surrounded by green trees of approximately the same age. These lightning-struck trees flamed out quickly, unwilling to spread fire due to either lack of ground fuel and wind or too much moisture saturating neighboring vegetation. Besides the cold, black single trees, I've found many larger charred areas during such outings. As I've scanned the grassy outlines that separate what burnt from adjacent green sage, I've wondered if these fires had ever been detected, let alone fought. Probably not.

Easier to miss, though even more common than the burnt snags, are the many lightning-struck trees that never conceived active flames—solid trees. Such trees contain little in the way of sheltered tinder, which is critical to birthing embers, and common in the old sentinel junipers whose cores are rotted out. Only in the event of fairly dry lightning will the duff skirting the bases of solid trees ignite. Otherwise, in a burst of energy, lightning gouges the tree's trunk from top to bottom, like the stripe on a candy cane. Strips of bark, splinters, and sometimes entire branches litter the ground. The scents of fresh cambium and ozone fill the air. Either way, burnt or not, these lightning-struck trees are something new and are a link to the past. Such vestiges from storms and seasons past aren't exactly what I'm looking for today, yet they are serendipitous finds that remind me of the irony in trekking across Rodman Rim: Regardless of what machinations and political contortions we go through, fires will start and burn and smolder until cooling ashes sink into the scabflats, whose soil once coaxed these now burnt hulks to reach toward heaven.

I'm glad I have the privilege of witnessing these connections inscribed on wood amidst an overabundance of stone, which continues to gnaw hungrily at my vehicle's tires. Still, the question remains: How many of the fires that we do extinguish would have gone out of their own accord? Seeing the limp-branched carcasses makes me wonder. Undoubtedly, the reasoning behind extinguishing even the most remote fire is to prevent a smudge from developing into a dangerous inferno—dangerous mainly in the damage it might inflict on human culture (for example, life, property, economic resources, aesthetic or recreational values). But in the total ecological and moral scheme of things, is black always ugly?

Could there be deeper aesthetic values realized only through risking the hot consumption of forest, range, and cultural artifacts?

I continue to snake the truck between a smattering of burnt junipers that never succumbed to chain saws and whose roots have never been exposed by pulaskis and shovels. To minds entranced by the management paradigms of "multiple use" and "sustained yield"—with all the commodity presuppositions built into these concepts—fire and rot are sheer waste. As fire perpetually reminds us, however, there is more to the world, and more value in the world, than meets the eye and the bottom line.

Surely, suppressing any and every single-tree fire should not be justified for the extra overtime pay it might yield. That overtime accrues from a remote, hard-to-reach smoke is a bonus to those whose sole income is drawn from lightning's residual flicker. But it strikes me as morally callous, or indifferent, not to see the irony in this situation: callous to the money spent to euthanize peacefully dying blazes; indifferent to the deeply historic and prosaic role of fires that run their natural courses in ecosystems accustomed to frequent lightning and periodic fires.

These are the things one thinks about while driving hour after hour on bladder-popping roads in search of a smoldering single-tree juniper fire that is most likely out. The chance of finding a two-day-old single-tree fire amidst one hundred square miles of nearly level juniper-studded terrain is not particularly good. So be it. Unfortunately, our helicopter is committed to other work, leaving roaming ground crews (namely, us) as the only spade to root out fires. Then again, this is fortunate, for we are given the opportunity to look and search, and cost-benefit analyses of our actions aside, searching for something new, something seasonally reborn though very ancient, is deeply meaningful.

I think there must be some prospector and historian in all firefighters. Prospectors of minerals are in search of something others want, for profit; they are also, I'd hazard a guess, drawn into their tunnels, sluice boxes, and overland ramblings in search of something new and possibly rare. Knowledge that no one else possesses can be dangerous; intimate and privileged knowledge—be it of a home, an heirloom, an artifact, a thought—can also lend itself to a sense of geographic and spiritual rootedness, to be shared with those who will likewise appreciate it. A broken piece of quartz crystal or a piece of obsidian worked by a human hand is present, but not all that common, on the edge I inhabit this day. These high-desert treasures may serve as geologic and cultural bonds to events and intentions leading up to the present. The scarcity of the geode or obsidian chip intensifies the link, possibly the beauty, and definitely the excitement of finding something new. Nevertheless, these assay out as bonuses to what we're ultimately after. Today's gem is smoke, if it can be found, a small ruby of a fire, hardly the mother lode of heat and flame and power embraced by popular culture. My culture. My past.

◆ ◆ ◆

Beneath a nearly translucent dome of parachute, a smokejumper, wearing what looks like bulky coveralls and a mesh-shrouded football helmet, descends to the ground through a dense forest canopy. There, by himself, he gathers his gear amidst a haze of smoke, then digs a small trench around a Roman candle–like snag. My throat is also dry, so I take a swig of milk and reach for another handful of Nilla Wafers. It's Sunday evening and I'm watching *The Wonderful World of Disney*.

I can't recall the name of that movie, or even the plot, but the image of that solitary firefighter frantically throwing shovelfuls of dirt onto a blaze is very clear. Such an image is also part of my youthful infatuation with employment ads in the back of outdoor magazines that announced CONSERVATION JOBS. These ads included tantalizing descriptions of the hardy values found in working out-of-doors: adventure, bronzed skin, freedom. I never did enroll in the correspondence courses that were the promised pathways to these jobs, since dropping out of the sixth grade was out of the question, and by the time I graduated from high school, college seemed a more respectable pathway for my desired career as a wildlife biologist.

So to say that my fascination with wildfire did not grow out of romantic, larger-than-life portrayals of firefighting and working in the woods would be false. I still enjoy watching the firefighter training films that at one time composed my catechism as a federal wildland firefighter. Though dripping with hyperbole and manly wrestling with nature, these films also defined how, for much of this century, those in the wildland fire service have conceived of themselves, what they did, and what they wanted to do. The films span the period from the late 1950s to the early 1970s. They portray fighting the fires of summer as a paramilitary, all-American, and heroic crusade against, for all intents and purposes, the devil himself. The firefighter might just as well be a knight riding into Jerusalem.

Hold That Line with Dirt, *Man against Fire*, and *Forest Fires and Water* are a few of the films that not only taught basic fire suppression skills but also advanced a distinct philosophy of fire and firefighting. There is a serious tone about them that makes their subject matter seem all the more urgent. They have a distinctive look and sound, closely resembling many of the nature films, travelogues, and newsreels produced during that same era, and earlier. Change the script a little, and you can just as easily hear a voice trying to sell war bonds. The melodramatic tone of the narrators makes fire suppression seem larger than life, which in a way it is, but not necessarily in the way the scripts read. It must have been great fun for whoever wrote the screenplays for these short programs, brainstorming ways to teach the troops and spread the gospel of conquering wildfire.

Take, for instance, *Man against Fire*. Against the backdrop of smokejumpers attacking a small blaze in the Deschutes National Forest of Central Oregon, the narrator frames the issue of fighting forest fires: "Man against fire—add blazing timber and mid-summer heat and you're at the hinges of hell. But wildfire can often be tamed before it gets out of hand. Luck has something to do with it: Land lies flat, wind is down, man has a fighting chance to beat the devil to the draw."

Fire personified. Like a legion of demons, "embers can't be trusted till they're dead out." Doug Maxwell, the fire boss on the Canyon Creek Fire of August 1968, and narrator for part of the film, makes no bones about the analogy between a campaign fire and war: "It's a mighty battle of men and machines." And at the point in the film when the Canyon Creek blaze purportedly gains the upper hand, the fire's status becomes "fire against man." When the firefighters' luck turns for the better, the fire is not simply contained, it is "strangled."

The list of catchy, anthropomorphized (usually sexist) phrases could go on. But even if the writers were just speaking figuratively, the consistency in metaphors reflects an evangelic zeal for an underlying philosophy, one that views the romantic value of fighting fire in a Jamesian sense of redirecting atavistic tendencies, which is to say, not in the value of fire itself. Oh, "the blood and sweat and grief of some thousand men. The strategy—they did the job." Manly men against fire: Smokejumpers are "rugged and ready ... dedicated, stalwart men, guardians of our forests." They are "earnest men" who "saw their duty, and a forest lives on; a forest where timberlands continue to grow, waters trickle and cascade, wildlife runs free and nests high—yes, a forest lives on, thanks to determined, courageous men who put out a fire in good time." The underlying supposition is that green is good and black is bad. In a sort of platonic "form of the green," a static, climaxed image of uncharred forest lands becomes the floral analogue of the beatific vision.

The narrator in *Man against Fire* even muses that "someday man might be able to prevent lightning-caused fires." At another time in history, it probably seemed equally compelling that we might one day rid the world of all predators. (If the lion won't lie down with the lamb, then the quicker we dispatch the lions, the better.) I'm sure that to those who viewed fire as the joint enemy of nature and humanity, eradicating its natural causes was an admirable goal: If we can just pull a technological reversal of cloud seeding, we could be rid of cumulonimbus's offspring once and for all. However, the real question for us today is, Is fire the enemy? If it is, then let's rid the world of lightning and wildfire, and once the task of our conscription against fire is complete, let's move on to suck in new hardihood, flexing our manliness over other appendages of nature. Then, let's sigh wistfully, in unison, with the narrator's final words in *Man against Fire*: "You know, it's good to see green again."

✦ ✦ ✦

Ah, sweet romanticism. Engaging fire can be romantic insofar as it encompasses an array of values, tapping into areas of the human soul that less raw encounters with culture and nature are hard-pressed to awaken. This can be a very good thing, seizing on the value of that which is immediate and unmediated. What strikes me, though, as quaint, a bit dated, even false, about the old training movies is that they breathe a sort of unbalanced romanticism that looks at engaging fire through rose- (or, better yet, smoke-) colored glasses. Wildland firefighting is a quintessential all-American endeavor, a wholesome outlet for subduing the earth; and fire is evil—a destroyer of life and property, a scourge of the earth.

A romanticized view of engaging fire has little room for both rusted steel spikes and small Gambel oaks growing on the same hillside under a hot August sun. The hero's journey may include both; if it reflects reality, it must.

I'm not sure if it's because we've grown accustomed to watching movies of epic proportions accompanied by powerful and dramatic melodies, but the wildfires of summer entice firefighters to view themselves writ large in Technicolor. For instance, in *Red Skies of Montana*, every time Cliff Mason (Richard Widmark) bails out of his jump plane, we are treated to the sound of a boisterous march. Similarly, the military-film genre is not without its examples of this accompaniment phenomenon. A classic scene in the movie *Apocalypse Now* shows the air cavalry in formation over the Southeast Asian landscape, each helicopter equipped with its own loudspeaker blasting operatic strains of Wagner. There's a certain perversity to it all. As opposed to *Red Skies of Montana*, in *Apocalypse Now* it's not just a matter of watching a film scored with music; rather, the event itself is scored. The combatants are both actors and observers of their own actions. We are watching an enactment of soldiers watching themselves, adding drama to their own actions, acting as if they are acting. Reality is objectified, and the moral complexity of strafing human flesh with a whirling Vulcan cannon, in time with music, is flattened out into a grand, real-time image.

But whether flying in formation over Vietnam, making a final approach for a retardant drop in a DC-6, or stalking a fire in a four-wheel-drive truck, there is the similar urge to dramatize the drama and replicate in reality what has moved us in the theater. If you've ever heard a fire crew leader bark to the troops, "Let's rock and roll," pealing rubber to the sound of sirens and the flash of rotating lights, you probably know what I'm talking about. We like to see ourselves cloaked in grandeur, often when our actions are less than grand. Nevertheless, it's hard to be both actor and spectator when attacking a fire, to choreograph and go out of our way to augment what is occurring before our eyes. When we do, we tend to overestimate our importance and, as a result, lose track of what we're trying to accomplish in the first place. Still we try. For there is a moodiness—and excitement—in the air while approaching, sizing up, and engaging a fire. This mood refracts the landscape, the atmosphere, and our own ambivalence toward the energy radiating out of grass and brush. To the extent that music can mirror the soul, to view such a scene does evoke melody.

✦ ✦ ✦

"Kickin' Country for Central Oregon—K-R-C-O," the mild Southern drawl announces over our FM radio.

Midafternoon arrives, and arrives, amazingly, without any flat tires. Tim and I have traveled only five miles, but it took us three hours; this is five miles we'd rather not retrace today. We scan our BLM quadrangle map, discovering that, to the southeast, we can reach another county road in about a mile. However, to get there, we have to negotiate through a sliver of private property. No big deal.

To enter the private ground, we undo a couple of fence staples that fasten a loop of barbed wire around the top of a kindling-dry gatepost. A foot away, a faded plastic sign reads NO TRESPASSING. Surely the owners won't mind. Official business and all. U.S. Department of the Interior. Bureau of Land Management. Engine 5560. Through the gate and having repaired our damage, we creep down the road—a road that gets better by the moment.

Tim glances over at me, then ahead. "Uh oh," he mumbles.

"Oh, shit!" I say, finally noticing that the road weaves right through the ostensible owner's backyard. "No getting out of this one."

"Maybe no one is home?" Tim queries.

There's a cluster of shacks—homemade looking, walls paneled with warped and water-stained plywood. We pass a rusted-out Impala that seems to be propping up one of the shacks, possibly a chicken coop. A white-enamel fence of Maytags and Whirlpools lines the driveway on each side.

The first sign of life is a blue heeler, who surfaces out of the roadside sage like a porpoise. It keeps pace off the port fender, barking all the way. Two men emerge from behind a white mobile home that sports a partially finished gable roof. Freemen? I wonder. I don't see any firearms, though I can't help but think the two probably don't take kindly to two G-men sauntering through their yard in a big yellow one-ton.

I ease the engine to a stop so as not to bathe the men in dust and make a bad situation worse. "Man the radio while I take care of the ass chewing," I tell Tim, as I start to get out of the cab.

"Howdy," I say from a distance. A balding, potbellied man, probably in his late sixties, balances himself against a red, mud-caked ATV. Next to him stands a younger man, thirty-something. The younger man's teeth are tobacco stained, chinked with what looks like Cheese Whiz. I try not to stare. I decide to make a pre-emptive strike by telling them we've been looking for a reported fire. The truth.

"Yeah, was one hell of a storm a couple nights back," says the older man. "My son and I went up on Silver Ridge thinking we smelled something burning. Didn't see nothin'. But get a day or two of hot weather, and the next thing you know the bastard will be off and running."

We exchange more small talk, and the older man recounts the time that some greenhorn BLM archeologist, or maybe it was a range technician, got stuck up the road after he warned her that road gets damned greasy when wet.

"She should've listened," he says. "Nearly got our own rig stuck trying to pull her out. Finally took a wrecker. Came all the way from Bend."

"Is that right?" I chime in. What can I say? I give the older man our dispatch center's phone number and tell him to call if he sees any smoke. He thanks me, though he says he'll need to drive into Brothers to use their radio phone.

It's funny how a perceived enemy—like fire—can soften borders. Funny in retrospect.

So I put our truck in gear and sheepishly continue on down their driveway, then onto a graveled county road, which after five miles lands us in the one-restaurant, one–gas pump town of Brothers, and then back to Prineville.

Fall

8

Dispersing

Initiate all action based on
current and expected fire behavior.

—STANDARD FIRE ORDER #2,
Fireline Handbook

Change is in the air. At first, only the perceptive notice it—
mourning doves filling their craws on gravel roads, the passing
comment about the beginning of classes. Summer still holds the
balance of power, but a new season approaches, and through it
all we continue to engage fire, even if those participating in the
engagement and the persona by which fire is recognized will
change.

By the end of August, a few students on my crew are packing
their gear and heading to the district headquarters to sign out for
the season. Among the curses for firefighters who attend schools
on the semester system is dismissal a month and a half before the
chlorophyll begins draining out of cheat, and registration while
fire is still full of shimmering virility. These emigrants retrace the
trails on which they struck west in the spring, and they do so with
smoke in their rearview mirrors. Only upon reaching Eastern

Colorado flatlands, by means of dangerously overloaded compact cars or contrail-stringing commercial jets, do the apparition and the reality of this season's smoke dissipate.

This can be a time of melancholy, both for those leaving and for those left behind. The thought of wildfires haunts these firefighters-cum-students in ivy-ensconced libraries as newspapers report free-ranging fires in the hardwood litter of the Berkshires, Adirondacks, or Appalachians, token fires compared to what blazed in the West. They are fires, nonetheless, and where's a damn pulaski when you need it?

For the lucky few enrolled on the quarter system, or for those who are committed till layoff time, fire is in the full height of maturity and a still-present actuality. Some call the month "dirty August." Heat can be intense, days are still fairly long, fires are usually plentiful, and legs are crusty black with dirt and ash. Grass and brush are fully cured now, and dryness has flowed up the flanks of ranges throughout the Intermountain West. Fire has also flowed to the lower latitudes, leaving the Yukon and Kuskokwim, in search of the Columbia, Snake, Missouri, and Tuolomne.

Nevertheless, late August in the high desert signals the end of lightning. If mature storms do roll in, they bring ever more blackness and ground-sweeping rain to bathe the sage steppe. Run-of-the-mill cumuli try to build the way they have since spring, but either they are blown apart by wind shear or whatever convective energy they initially wield quickly tires and they feather out into opaque, wispy clouds. Like inverted pyramids, they furrow north, trailing meager virga and little hope of further buildup; pay periods now mark forty-hour weeks. The ocean-born pinwheel that first ushered in spring fire only occasionally appears; now, fires are mainly the result of human negligence or malevolence. And amidst it all, the elk begin bugling.

✦ ✦ ✦

To engage free-burning fire is to dance with a force of nature that has defined hominid existence from the labyrinth of prehistory. To hunt animals is likewise to engage the rhythms and cultural realities that stretch back to our ancestral quest to survive. So in the midst of dirty August, between fire assignments, I fling arrows at a cotton-stuffed target butt in our guard station compound. The critical question this summer, as always, is whether or not my archery tag will go unused while I choke down dry sandwiches of ham on white bread along some hot fireline. I hope that rather than backburning more acres I'll soon be burning my meager annual leave hours as I creep around in the sage and pine with olive and black paint smeared on my face.

The first year that I bought an archery elk tag, I found myself mid-September in a sprawling fire camp near the community of Buck Meadows, California, just west of Yosemite National Park. It had been the sort of road trip that all engine personnel long for: good overtime, massive burnouts, multiple convection columns drawing their thermonuclear-like plumes into picture-perfect unity, and homes of noted celebrities to protect. But our assignments were starting to get old. Nights were cold. Inversions, which blanketed camp in a preternatural haze, drew out longer in the mornings. Oil-fueled smudge pots and crude oil–like coffee failed to thaw out the 5:00 A.M. stupor before the morning briefing. The dew was getting heavier. And our strike team was now committed to line rehabilitation work that amounted to cutting and stacking brush and seemed unrelated to any damage caused by the suppression effort.

Plus there were only two weeks remaining in Oregon's archery elk hunt.

So I finagled an escape from camp with a crew member who had to be demobilized to go back to school. In a rare instance of demobing expediency, and with completed time reports in hand, we were soon on our way—up over Tioga Pass, north on U.S.

Highway 395, and a dozen or so hours later (having exchanged the piñon-juniper landscape for juniper and bitterbrush) in the familiar high desert country of Central Oregon. The next day I would be packing my red 1970 Datsun shortbed en route to the Ochoco National Forest, to the elk I hoped would be rutting. There is a time for fire, and there is a time for hunting; there is also a time for hunting amidst fire.

On years that I hope to bow hunt, I also hope that late August and September will deliver cool temperatures and dampening rains. Such paradoxical hopes for a firefighter are spun out of the conviction that the chill in the air will prompt the bulls to become a little hornier and more vocal. My other concerns are with meat spoilage and my own comfort. To hope for cool temperatures and precipitation causes me no little anxiety, though, for September is still a time for fire.

My ambivalence and my schizophrenia about hunting during (wild)fire season are quite evident when I continue to carry my two-way radio on such outings, just in case things back home get really serious and they need an extra hand. There are times that I stray out of radio range, but there is something comforting in having the couple extra pounds of electronics along—just in case. Misery loves company, and my fire envy gets the best of me when I wish that no one back home gets on a fire while I'm out hunting. I don't want to miss any action, I think to myself, as if one more fire will grant my peers (and deprive me of) some life-changing fire experience. However selfish that may be, part of me desires nothing but project work for my comrades on the home front.

With bow in hand, I continue to watch the clouds, both to perceive the air currents that might signal my whereabouts to incoming bulls, and to sight late-season cumulis creeping into the southern half of our district. I carry a pulaski in my truck in the event I come across a fire.

Like our fishing trips of spring, a number of these September

expeditions blend into one another, for the contours of their progression are similar, as is the topography upon which they unfold.
The last evening of the hunt is usually spent sitting around a campfire with a fellow nimrod or two. There is plenty of dry wood handy,
and during the warm September afternoons mountain mahogany
graces us with shade. At night, dead mahogany branches melt
down into bright-orange coals, within which we bake potatoes and
upon which we brew coffee. The mahogany smells sweet, and it
burns hot. The landscape is still dry enough that our campfire
could get away if we aren't careful. We'd never hear the end of
that. Typically, we bag no elk, though we see them. Quite a few
answer our bugles. This is success.

In the bugles and chirps of September elk, a deeper metaphor
and mythology are recapitulated: fire and sex. (Bachelard was apparently not an elk hunter, for I'm sure if he had been, his psychoanalysis of fire would have mentioned the rut.) Early on in the season, like a troop of fur-robed Boy Scouts trying to start fires by
friction, the bulls thrash their itchy, velvet-covered antlers on
brush and small trees. And as the sweet fragrance of cow elk fills
the September air, the bulls grunt and whistle and piss to establish
territory and flaunt their dominance. They do this when fire is at
its peak; they continue to do this when fire begins to wane. In the
shadow of fire and mountain mahogany many a Western archer
sneaks through the wildlands to glimpse the yellow-tan butts of
these great hoofed animals. For those of us who fight fire, once
the hunt is over, we return to our tool caches and duty stations renewed and ready to chase the few smokes that will continue to
crop up while there is still time.

✦   ✦   ✦

The fire of spring is fresh, newborn; the fire of summer is adult.
Come September, fire turns silver. It is not yet enfeebled, for it
still has the potential to blossom in the bone-dry vegetation, but

its days are numbered. Even if the fine grassy fuels that nour-
ished fires all summer remain desiccated into November, the di-
minished daylight and increasing dew, then frost, skimming the
ground at night, all crimp the extravagance of free-burning fire.

Seasonal strength-of-force rosters have been stripped by college registration lines, leaving skeleton crews spread ever more
thinly over vast acres of terrain. And since the fires that do burn
oftentimes collapse onto themselves under the cover of high
nighttime humidity, some of the urgency in jumping on them
quickly and keeping them small likewise evaporates from the en-
terprise. This allows time for contemplation—contemplation of
why I'm still out here on the fireline while others have quit to go
back to school; of whether this is a real job; of how this connects
with other things that I find important in life; of how I can walk
through the district office summer after summer and still be
asked by receptionists and administrators, "May I help you?" (as
if I'm some visitor wanting to buy a woodcutting permit); of what
significance this is, and of whether I even care. And what does all
this mean to a spouse who experiences only secondhand the ash
and embers and stars blinking through a calm October inversion?

For those on fire crews whose primary duty is initial attack, the
solitude and contemplation afforded by elder fire is what crys-
talizes the engagement of wildfire into something more than a
necessary evil. There are no greasy catering trailers to belly up to
or paper-shuffling demobing hoops to jump through. There is the
freedom found in scratching around in a fire with a small cadre of
smoke eaters, friends, who want to extend summer into fall.

✦ ✦ ✦

Even while wildfires are still being extinguished, other fires are
being set at an accelerated pace. Like spring, fall is a traditional
time for prescribed burning. But unlike spring, many of the fall
burns—at least on the sage- and juniper-covered highlands—are

large. A burn of several thousand acres is not uncommon. In areas
of sparse ground fuels and flat topography like the Millican Valley,
conditions must be severe to propagate fire: Low humidity and
high wind are needed to carry fire, which otherwise would merely
settle into individual bunches of grass and sage, separated by several feet of open dirt and pebbly anthills.

Moreover, like the prescribed fire of spring, prescribed fire during the fall is weighted with moral freight. Over the past century
there has been incessant debate about firing the woods, prairies,
and wildlands, and this for a whole host of reasons. Whether
called "light burning," "controlled burning," "management-ignited
fire," or "prescribed burning," the intent remains steady: Introduce fire for some intended good. The pendulum of scholarly and
popular opinion on whether to burn or not to burn has swung
back and forth. Today, the advocates of burning seem to have the
momentum, with the validation of historical precedent and scientific research on their side. This, however, is not without its con-
troversy. Even if we grant that anthropogenic, or human-caused,
fire has played a decisive role in how ecosystems and landscapes
around the world have been shaped, and even if we grant that because of this historic relationship between humanity and fire it is
impossible to reclaim a golden age of natural fire-ecosystemic balance that has not existed for millennia (if ever), we are still faced
with intractable moral questions about how to use fire.

Many questions need brutally honest appraisal: What sort of
landscapes do we want? Are we enamored of a pre–European settlement image of forest "health" that existed only because American Indian fire practices brought a certain look to the land?[1] If so,
why is that good? And what role do these cultural fire practices
play today? Do we want to increase the wildlife-carrying capacity
of our wildlands, and, if so, do we realize that this is a cultural de-
cision not inherent to or required of ecological systems? Are the
species of wildlife that fire is being used to promote native or ex-

otic? We want to reintroduce fire to decrease fuel loading and the potential for disease and so-called catastrophic fire, but are those things bad, per se, for the integrity of the ecosystems we wish to maintain? And for whom is ecosystem integrity a good? If fire is introduced to replicate a pre-Columbian landscape (if this is possible), and if one of the intended benefits of doing so is the replacement of water-hungry shrubs and trees with soil-stabilizing grasses, why is that good?[2]

What all this boils down to is not so much a matter of science—as important as that is for ascertaining causes and effects of certain actions or inactions—as a matter of philosophy, a matter of values and ethics, and a matter of what is truly important to humans in a world of which we, by virtue of our ability to wield power, are stewards. No multiplication of charts, tables, or arcane scholarly monographs will definitively solve the questions of how we ought to use fire. This is not to say that there is an impenetrable barrier between facts and values, for such a dichotomy is false: How we decide to interact with fire and earth is intimately bound to what fire does to earth; it is also intimately bound to what may be aesthetically and morally good, in and of itself. These are issues that we need to discuss, debate, and identify for what they truly are—moral.

It is quite appropriate that people are skeptical of a government that for many decades preached Smokey Bear rhetoric with religious zeal, a government that is now beginning to propound a doctrine more charitable toward fire.[3] Anyway, not everyone embraced Smokey over the past half century. And many of those who did, did so because they wanted to embrace something like a friendly bear at war with fire. Undoubtedly, the truth will lie somewhere in between loving fire and hating fire. That truth will probably attest every bit as much to who we are and who we want to become as to what is best for the flourishing of the living world in which we dwell. Maybe it can reflect both.

✦ ✦ ✦

While meditating on these things amidst the lingering intensities of elder fire, those of us still on the strength-of-force roster continue to—oh so imperfectly—light fire. We do so with the same mixed emotions and moral ambivalence that we felt in backfiring—engaging the flaming fronts that rushed up toward golden wheat fields between the Deschutes and John Day Rivers—or the ambivalence of looking at trees that burnt out in a silent death under the darkness of flashing thunderheads. Regardless, it's the time of year to burn, and we get paid for the burning.

Like life, fire evokes comedy and tragedy; it distills them, like an alchemist's hope of producing a refined ingot of gold out of common elements. But instead of drawing gold out of base metals, fire draws comedy and tragedy out of sage bark and juniper duff and red-needled logging slash.

A moment that epitomized the comic absurdity of what we sometimes do when we unleash fire occurred when I worked for the U.S. Forest Service one fall, during a clear-cut burn. I much preferred open-range burns to the stuffy little broadcast units that we kindled in the forest. There was usually far too much smoke from the clear-cuts for my taste. I think that I sucked in more smoke during that one fall than in all the years, combined, that I've engaged wildfires. For the most part, this was because I was invariably a member of the holding crew. My job was to either stay with a fire engine located on a road or cat trail or position myself somewhere along a pencil-thin hand line running up the side of the burn unit. In both cases I was supposed to stand guard and hold the fire from going outside its prescribed boundary. Of course, the greatest potential for an escape, or spot fire, is downwind and downsmoke, exactly where the holding crew was positioned, meaning me, the "smoke follows beauty" adage on a grand scale.

One particular day we lit off a unit composed of about five acres of medium-heavy fir and pine slash. I can't recall whether or not we plumbed the unit with a hose lay (composed of a primary "trunk" hose from which secondary lateral hose lines would branch) to wet down the adjoining forest litter as a pretreatment against spot fires. It didn't make much difference, for when the wind laid the smoke over the fireline (where I stood with a shovel), we started picking up numerous little smokes in the forest duff. Punky logs looked like they had contracted some form of deadly contagion as they developed black smoking spots all over their powdery, reddish brown surfaces. Unless you jumped on the malignant spots with hand tools and excised them quickly, the whole log would soon be a mass of flame and a pain in the ass to mop up. We made good headway on catching all the small spots.

Then the wind sheared close to the ground, driving an impenetrable screen of smoke at eye level. I pulled my red bandanna up over my mouth and nose, but it gave little relief from the acrid fog that by now was burning my eyes so badly I could hardly see through the streaming tears.

To hell with this, I thought, as I started feeling my way up the bare-dirt fire trail to the road above, where I hoped to escape the smoke. By the time I reached the top, I was gagging and coughing and crawling on my hands and knees. Finally, I was in clean air and surrounded by the clean, yellow Nomex of engine crews gazing down into the broiling smoke.

Before we started the burn, I noticed that one of the timber guys (some silviculturist who was enlisted—like many office folk during prescribed burns—to lend a hand in a supervisory role) was wearing some sort of gas mask. He was wearing it with the swaggering pride of someone pulled from his official duties to supervise people who work in fire all the time. He had been running up and down the line, screaming at people to jump on this or that smoke, at least that is what it sounded like he was saying through the black rubber mask. During my infantlike traverse out of my holding zone, on hands and knees, I lost sight of the holding boss in his anti–mustard gas mask.

Then he appeared. He crawled up over the lip of the slope and onto the road. He tore the mask from his flushed red face, only to reveal snot rolling down his upper lip and around his mouth; tears flowed profusely from his black-pitted, cherry-red eyes. Mr. Macho Gas-Mask Man, I thought. He still frantically screamed about spots, and about the fire getting away. As I looked into his mucous-covered face, then at the charcoal-filtered mask hanging limp in his hand, all I could do was smile and look away.

There was only a minor slop-over that didn't take long to mop up. Other than breathing far too much smoke, no one (and nothing) was worse for the wear.

Maybe it's one of those situations that requires your presence to see its humor, but comedy there was—an absurd sort of comedy that blended overreaction, blustering authority, and proud dependence on technology that seemingly delivered more smoke to its host than it dispelled. The entire scene was a reminder not to take ourselves too seriously, even while asking difficult questions about the way we use fire.

✦ ✦ ✦

By mid-September, the fires that still break out over the Columbia River basalt of North Central Oregon are more an anomaly than a concern. Smoke columns that a month earlier would have aroused frantic 911 calls and recon flights by our district's helicopter are now viewed with the knowledge that Sherman County farmers are burning wheat stubble. Fires that start at the bottom of the canyon—from hot railcar brake shoes or river runners who think that the environmentally correct thing to do is burn their toilet paper—may race toward the canyon rims during the heat of the day, only to find fallow ground, or fields long since harvested.

Oftentimes they never make it that far, as the early-fall humidity recovery is enough to drive the creeping fire into a black margin of cooling ash and smoking sagebrush stumps, midslope.

This time of year, we close our northernmost guard station and relocate personnel to stations closer to district forested lands, like Paulina or Dayville—areas where a fire can gasp through the night in heavy duff and reawaken to adulthood during the next burning period. It can be a sad time, making the final patrol of the season along the Deschutes River access road, realizing that one might not see the yellow and sage-green slopes of the canyon walls for another six months. Also, when it comes to seasonal work such as fire engagement, you can never be too sure where you'll be the following summer; it might be your last patrol. Such a sojourn is rarely necessitated, other than as a way of bringing closure to the byways and geographic fixtures that anchor one to a region of fire.

I drive upriver as far as I can. The road dead-ends at North Junction, the spot where, around the turn of the century, the rivaling Hill and Harriman railroad magnates, who had doggedly raced their lines up opposite sides of the Deschutes River canyon, decided to join a common bit of rail out of sheer pragmatism (and to avoid right-of-way conflicts). The tracks on the west side of the river have endured and are now operated by Burlington Northern & Santa Fe Railway. The washboard access road on which I drive is the eastern railbed of the Harriman interests, abandoned in 1935. How times and fates change.[4]

There is still plenty of good fishing going on for resident trout and the few summer steelhead that dodged nets and dams to make it this far. Some insects will continue to hatch all winter, and the red-sided rainbows will continue to rise.

I turn around at North Junction with the sun dipping beneath the Mutton Mountains to my west. The Muttons anchor the northeastern boundary of the Warm Springs Indian Reservation, and their exposed eastern flanks spill stringers of ponderosa pine and bunchgrass down to the Deschutes River. This is a margin formed by a treaty. It hadn't always been so.

Quite a few seasons end this way, and the ritual remains the same no matter what section of river hosts the final patrol. It's not the end of fire, but the end of a season of fire and a tangible border between summer and fall. Looked at with the elk in mind, it's just another transition point in a season that has already broken in, even if, until now, we resisted its clammy presence. Now it cannot be denied any longer; it's time to move on. This gold-baked land of grazing, fire, and public domain is easily sentimentalized in the realization that it won't be seen for an indeterminate period of time. The reality, however, is that it also included many frustrations: excruciatingly hot days that hardly needed fire to make you feel like you were at the hinges of hell; blown sidewalls on inside duals—wounds inflicted by the rugged access road; obnoxious urban tourists whose ivory-white beer bellies hang out over tight bikini briefs; fires that drag you to the brink of heat exhaustion.

But in the contemplative quiet of fall, after the Labor Day rush and in the soft glow of bunchgrass aflame with the last rays of a harvest sun, other sentiments dominate—farewells, regrets, hopes, and desires to rekindle relationships strained by distance and time, relationships tested by the heated dervish of fire and ash and pulaski and shovel.

With the first freeze, most of the doves head south, even if the harvested grain fields of the Columbia Plateau still provide feed in abundance. With the station winterized, we ascend out of the canyon and, likewise, head south.

9

Turning

Shot gold, maroon and violet, dazzling silver, emerald, fawn,
The earth's whole amplitude and Nature's multiform power
    consign'd for once to colors;
The light, the general air possess'd by them—colors till now
    unknown,
No limit, confine—not the Western sky alone—the high
    meridian—North, South, all,
Pure luminous color fighting the silent shadows to the last.

—WALT WHITMAN, "A Prairie Sunset"

All the world is gold. Everywhere you gaze, the landscape looks like sunbaked straw. Cheat and fescue and rye are about as dead and dry as they will get until fall rains soak into their stems and winter snows blanket their crowns. Tans and grays and greens also swell and ebb across the Intermountain West: Sage, bitterbrush, juniper, and, higher in elevation, pine and fir add their own portions of life-giving color. But from the Columbia Basin south into the Great Basin, everything is welded by and frosted in dull gold.

By the beginning of October, the first rifle crack of buck deer season has echoed against Oregon's desert rimrock, and my fire crew now roams back roads on hunter patrol. Such patrols are always ambiguous missions, even if we all know the routine well. Crews are thin this time of year, so everyone even vaguely associated with fire takes an active part. The helicopter contract has expired, so birdless helitack members team up with what few engine personnel remain on duty; fire prevention and fuels specialists (and even the fire management officer) all take to the road to make a presence. In theory, we're supposed to give hunters cause to think twice about carelessness with fire.

As if on a religious canvas of the neighborhood, each engine crew is assigned a patrol province: Glass Butte/Hampton, Bear Creek/Millican, the Badlands, the Mauries, the South Fork of the John Day, the lower Deschutes River. With standing orders in hand and miscellaneous Smokey paraphernalia on board—garbage sacks, coloring books, pencils, rulers, and so on—we fan out.

We have no real authority. We can politely tell people to put out their campfires if burning bans are still in place. They can tell us to go to hell if they want to, and then it's our prerogative to either bid them farewell or call for a federal ranger to intervene with a citation for the reprobate. Knowing that it can get ass-freezing chilly if you're camping, we sometimes just keep driving.

Given the meager number of escaped campfires in our patrol regions in recent years, such excursions might seem to be lingering anachronisms. True, our presence is known to the smattering of camps and road hunters that we pass, and, I suppose, we are more or less in position in the unlikely event that a campfire or cigarette butt gets out of hand. But whatever their efficacy in serving as deterrents or ratcheting up our state of suppression readiness, these excursions do serve a legitimate—if somewhat selfish—purpose: They are a sanctioned opportunity to explore new country. New access routes to old fire haunts are driven; old routes to new destinations are checked out. In rambling over this country's lava flows and crested-wheat seedings, we learn about a history shaped and informed by fire. We see where fire is most likely to be born, where it might in the future be corralled, and where we might escape in the event that the fire's intensity threatens our lives. Much of this is absorbed in a sort of unthematic, unconscious way. Yet, it all goes into making us more informed firepersons, and persons informed by fire.

✦ ✦ ✦

Midway through deer season, there is a report of a smoke on a ridge about five miles west of the G.I. Ranch, an expansive spread located where the headwaters of the South Fork of the Crooked River bubble up in a high desert basin, just southeast of the geographic center of Oregon. The fire is most likely a holdover from a recent lightning storm that, by October, is very much an anomaly at this latitude of the high desert. The fire presents little urgency, even though we desperately want to take action on it, since it could very well be the last fire (and hazard pay) of the season.

So Pat, Robert, and I cram into the cab of my engine and embark on a night out in the field. Unlike fighting structure fires, where every second of response time may make the difference between life and death or between mere smoke damage and a gutted building, response time to many of our wildland fires is measured in hours. If it's a single-tree juniper fire that you're dispatched to, it's not uncommon that by the time you get there it will have burnt itself into a smoldering column of charcoal and smoking duff. And if the tree isn't totally consumed, all that means is more mop-up—mop-up that separates green from black branches and requires constructing a large enough boneyard (ground scraped to bare soil or within a cleared area of black) to hold all the charred debris, to avoid an embarrassing rekindle.

Homesteaders nearly a century ago, lured by enthusiastically worded ads, attempted to make a living in the area to which we are headed. Some succeeded; others, for a time, barely subsisted in the semiarid sagebrush steppe and eventually capitulated to the elements and moved on to more verdant pastures, leaving their homes to the sage rats, magpies, bats, and weather. Some ranches remain, like the G.I., and many of these lease grazing rights on adjoining BLM and national forest lands. For the most part, though, today's grubstakers are the sage and juniper, the coyote, the elk, and the deer that move across the sandy and obsidian-chipped plateaus and playa.

Even earlier than the settlers, this land was visited by an 1845 wagon train en route to the Willamette Valley, led by Stephen Meek. The train was taking a purported shortcut when it became lost and wandered in the desert for many days in search of water. Journals tell of how the train often traveled at night, navigating by fiery cairns of flaming sage set by advance scouts. All this route did was cut short the lives of many of the emigrants.[1]

Allegedly, one group in the train stumbled onto a creek bed in which they found a few gold nuggets—a spot later dubbed the Blue Bucket Mine. As near as I can tell from the sketchy historic records of the wandering train and from the legal description given by our dispatcher, the area in which our fire was reported is somewhere in the vicinity of the lost gold. It's funny how such magnificent finds are lost. Maybe it's not so unexpected, as time blurs the margins of hope and reality.

Three hours after leaving our fire cache, and with a fair amount of wandering across rarely used roads, we catch sight of what appears to be low-lying haze on a ridge called Scammon Butte. Rolling down the window and letting the fall breeze penetrate the cab confirm it—smoke. Just as suspected, it turns out to be a single juniper tree that has obviously been burning for a good while, for much of it has already turned to ash. Since night is quickly approaching—and, with it, mid-October temperatures—we see no reason to waste a good fire to premature drowning. We call Dispatch to inform them of our discovery and give them a corrected legal description. They give us a fire number for our incident, which we'll need for our time reports. Then we get to work.

We scratch a line around the base of the tree, encircling the blackened area of burnt duff and grass and brush that extends ten or so feet out from the trunk. We cool the base of the tree with just enough water to allow us to sidle up to the trunk and cut off what limbs and branches had fallen over from the gently flaming hollow core. I take the plastic guard off the chain saw's bar and make sure the saw is filled with gas and bar oil. The chain looks sharp enough for what little saw work the evening holds. Some branches still have green needles and blue berries on them even though the bark is scorched. We heap these branches over the flaming core of the stump, and there they simmer. The sappy green needles flare up quickly, soon leveling off into a constant flame. We hold some of the branches and slabs of charcoal-fleshed wood in reserve to throw on the fire a little later in the evening, as temperatures continue to chill and our appetites grow. There's still cooking to do.

Years ago, we had an old fire dog named Walt on the crew. He prided himself on his culinary expertise on the fireline, back in the days when C-rations were standard fare. Walt loved to bestow such wisdom on the rookies as the intricacies of balancing little green cans of gelatinized morsels on the engine manifold, or the proper way to mix the components of various B units into a fire-line smorgasbord fit for any smoke eater or jarhead. As for me, I think there are only so many ways to garnish C-rations. Honestly, I don't think they tasted all that good. It helped if you were really hungry. But I do sort of miss them. I still have one of the U.S. Speaker can openers, which were included in the rations, dangling from my key chain; we don't get can openers in the freeze-dried, vacuum-packed MREs, or the name-brand boxed food rations that have pop-off lids. Of the C-rats, the B-3 unit was always my favorite. Among other things, within the small cardboard B-3 unit was a little foil-wrapped disc of candy that we called a John Wayne bar. These were supposed to be chocolate, I think, even though they had a waxy sheen to them. They were shot through with crunchy little tidbits, sort of like Almond Roca.

In the Forest Service–produced movie *Man against Fire*, the narrator tells us that "it's important for men to have their chow pipin' hot, refrigerator cold." I'm sure Walt would have agreed, although I'm not so sure Walt's concoctions are what the movie's makers had in mind, nor do I think that my fire supervisors over the years fully digested the import of this message.

On this night, sitting next to our single-tree fire, a fire of fall, we straddle two ages: the C-ration and the MRE. Almost all of our district's supply of C-rats has been consumed or disposed of, so MREs are pretty much all we carry, unless we buy our own rations. So in stubborn defiance, we pull out a few cans of store-bought chili and disregard the government-issued transition foisted upon us. We open the can tops to let off pressure during heating, then push the cans down into the searing bed of coals still encircling the tree stump. From the back of our engine we pull out a couple of lawn chairs ("Initial Attack" stenciled on their backs) kept on board for just such an evening of repose and feasting. Sitting in our chairs, we eat our piping-hot food in the warmth of the lightning-kindled tree, drink lukewarm water out of orange gallon canteens, and watch the stars come out.

Finally, we prepare our bedrolls. One advantage of working on an engine is that there is always room to carry a cot. There is *always* room. So we unload three cots from the engine and set them up in a semicircle around the hot perimeter of the stump. The incinerated gray duff is soft and powdery. Digging into it with a shovel, however, reveals that its serene appearance is deceptive. An inch or so down, it's blistering hot and, when disturbed, rolls and cascades like a molten ash flow off a volcano's flanks—Vesuvius in miniature. Probably better not set the cot there. We place the cots as close as we can to the fire, straddling warm duff to provide a little extra heat for the yellow, meagerly insulated government bag that each of us sleeps in. I'm a little hesitant to employ this technique, for I'm not sure what the ignition point is for the flimsy, drab nylon that forms the cot's bed. What the hell.

Morning breaks to sleeping bags still suspended above an ash bed, for which we are thankful. Between us is a barely smoking rim of black—what's left of the stump. Not much is said as we each, lazily, roll up bags, fold cots, and snack on stale crackers and Spackle-like cheese left over from dinner. Not much needs to be said. This will be our last fire of the season.

No need to conserve water, either. The two hundred gallons that we freighted in will stay at the fire and are promptly delivered to the stump by way of a chrome inch-and-a-half nozzle, designed more for structure firefighting than anything, though we carry it for just such a rapturous deluge. Robert serves as nozzle man, blasting like some crazed hydraulic miner into what little hot ash remains. It is only fitting, after all; the Meek train did say that they found gold in these parts. But today, our gold would be found on our time sheets to the tune of time and three-quarters, in the cured grass within which our little spot of blackness hunkers, and in the value of a shared fire over which we suspended our slumbering frames.

✦　✦　✦

Even Indian summers eventually come to an end, and with the inevitable turning of summer into fall, and wildfire into the last remnants of prescribed fire, so, too, the season of wildfire in some individuals' lives contracts to a symbol of youth. Engaging wildfire is not a young person's endeavor, but a fit person's. But even for the fit, seasons change.

Some persons tire not of fire, but of the suffocating bureaucra-cy-encrusted government agencies entrusted with the flame. A point of diminishing returns is reached where what originally at-tracted one to the thin lines of mineral soil separating combus-tion from verdancy is outweighed by a system seemingly stacked against the seasonal employee. A bitter divide emerges, much as in a divorce, and a shadow is cast over the passion that once was, might still have been, and to some extent always will be.

Others leave the engagement of wildland fire in a bittersweet transition to the careers for which they have prepared their entire lives. Fire always punctuated semesters and quarters of educa-tion—being embraced as the happening that it is and for the en-gagement that it requires—but firefighting has always been recog-nized as a transient occupation. Fire served its initiatory role, but other callings and vocations supervene to draw such individuals into new seasons of life and love and mission.

Some of these who have moved on from the fires of summer eventually return, if for no other reason than to revisit fire, en-gaging it one more season before finally pushing on to other en-deavors; or it may be revisited until one is no longer physically able to hike its smoking flanks.

Then there are those who are unable to ever fully leave wild-fire. They may, of physical necessity, remove themselves to a sedentary scoping of the horizon for smoke while ensconced in a glass-encased lookout tower. Or they may move up through the fire organization to supervisory roles that—while removing them from direct contact with flames—keep them in contact with the world of wildland fire.

A few individuals fight fire for a season or two or three but never really move beyond seeing it as a sooty, laborious, and not-that-well-paid chore. Fire is something to play with, but the sinews that bind it to life and death, to history, time, and the eternal, seem like mere gibberish. Occasionally, one finds a person in this category who has spent much of his or her life in fire but would just as soon jump ship if only a better-paying job (or pension plan or early buy-out) could be found. I consider such a one an anomaly. At least I hope this is the case.

For most who have engaged wildfire (however obliquely), and regardless of what digressions their lives may take, the smell of smoke, the crack of distant thunder from billowing white clouds on a midsummer's night, or the report of a firefighter's untimely death, all evoke sentiments of nostalgia. A photo album, an old pair of boots, or the woodcraft of how to mop up a campfire, all hark back to the grandiose hearths measured in acres, not in square feet. For a rare few, the autumn of their lives will end in one final season of fire. Such was the fate of Aldo Leopold, who died from a heart attack at the age of sixty-one "while helping his neighbors fight a grass fire."[2]

Fire remains, even if the seasons change.

✦ ✦ ✦

The hues of autumn fire span the continent. In the Northeast, the world is afire with yellow and red. Leaf peepers flock by the busload to Massachusetts, Vermont, New Hampshire, and Maine to view the annual turning of the foliage. The sugar-rich maples of the White and Green Mountains throb like so many embers, a cool conflagration of luminescent bulbs rolling over the New England landscape. Once their spent leaves drop to the ground, the trees become tinder for a smattering of hot autumn fires fanned by Atlantic winds.

Two thousand miles to the west, between the cordilleras of the Rockies, Cascades, and Sierras, deciduous fires also burn in the angled October sun. Aspen, willow, cottonwood, and alder are all in stages of turning. While these lack the brilliant reds that mark their Eastern cousins, their contrasts are nonetheless striking. Groves of flaming yellow aspen bloom out of little alcoves in the pine and fir overstory. Their white bark stands stark against the darker conifers, a giant, dew-glazed bouquet of daffodils. I'm attracted to these pockets of gold and white. They mark seeps of spring water and moist soil; they shelter hoof-pocked elk wallows. The autumnal fires that radiate through their leaves intimate the biological reality that many species, like aspen, are rejuvenated by fire. In a subtle refraction of light on gold leaves, this is a metaphoric suggestion that links renewal to change.

◆ ◆ ◆

I suspect that many a youth's first experience with flames under open sky didn't come while standing around a campfire but in the smoldering heap of burning leaves under an autumn moon. My life with fire, too, was born of fall.

I spent my first eight years living in a white clapboard farmhouse in the northern Willamette Valley, between the towns of Canby and Molalla. The two-story house was built in the late 1800s and was surrounded by one hundred acres of pastureland and brushy fencerows hedged by blackberry thickets that attracted a handful of ring-necked pheasants and small songbirds. The brush rows were my forest, my jungle, my secret hideaway. We dubbed the place where we lived the Heinz place, after old Mrs. Heinz from whom we rented the house. Another neighbor leased the surrounding pastures, on which he grazed cattle. Even though we didn't earn a living off the farm, it was our home. The extent of our farming consisted of gardening and yard work. We had a few gigantic walnut trees, and under their outstretched limbs, on the lawn, we laid sheets of plastic to catch both walnuts and leaves. For what seemed like weeks, we gathered nuts, removing their slimy black husks, which ensconced wrinkled, woody interiors. We placed the stripped nuts on racks of window screen to dry, then either sold them whole or shelled them and ate or froze their crunchy gold meat.

But it was the shed leaves of autumn that fascinated me most. After raking them into small piles we would transport them—either by wheelbarrow or upon plastic sheeting—over to the garden to be burned. Living where we did, on the windward side of the Cascade Range, precipitation was abundant, so very often the leaves would require a good bit of coaxing to sustain combustion. Usually this equated to Dad's pouring lawnmower gas on the pile and flinging a match onto the heap. With a whoosh, the whole pile danced with flame, over which hovered a small mushroom cloud. Depending on how wet the leaves were, the initial flashover might be short-lived, leaving only a thin layer of blackened walnut leaves on the surface, and white smoke slithering out of the pile's gut. I was often given the job of standing guard with a pitchfork and garden hose, the pitchfork to stir and coax the meager flames into hot abeyance, the garden hose, presumably, to cool the blaze down if the flames started to march through the dried corn stalks and pumpkin vines of the garden. Amidst the omnipresent dampness of the Willamette Valley, an escaped leaf-pile fire was more an abstract possibility than a real concern, though I'm sure that keeping a hose handy was what Smokey would do.

One October, I convinced my mom to buy me a devil's costume for Halloween. At age six, I gave little thought to the religious implications of this masquerading. It was (and maybe still is) just a costume. In fact, since I had already acquired some attraction to fire—from burning leaves and incinerating blades of grass with Dad's magnifying glass—it only made sense to don such lurid garb; after all, the picture on the costume box showed flames. The suit itself was made out of bright-red flimsy nylon material that was slightly frayed around the bottom of the legs. The box also included a grimacing black-and-red mask with two goatlike horns projecting out of the head. Best of all, though, was the little four-pronged pitchfork that was sold as an accessory (and which I hounded Mom to buy for me), replete with a red plastic head attached to a two-foot-long black wooden dowel.

I'm not sure if my parents ever knew, but one day, while cloaked in my devilish outfit, I went out to our still-smoldering leaf pile and danced around in the glowing leaves. So as not to burn my plastic fork, I used sticks to jab in the fire, exposing the compactly burning interior of the pile. As I look back on it, this is little different from poking sticks into a campfire, or drawing in the dirt with a piece of bark or a boot toe. It's a way of nudging the environment and watching things move.

At the time, my costumed glorying in the burning leaf pile must have seemed to be all of a piece: My dad gave me a real pitchfork to tend the fire; the devil's costume included a pitchfork; ergo (by the turn of a six-year-old's syllogism), my costume and mock pitchfork belonged alongside the flames. A scene fit for Dante. Anyway, the devil didn't seem like such a bad guy since, to my thinking, fire was sort of neat, Halloween dealt with mysterious things, and fire was mysteriously hot. It's a wonder that those frayed nylon pants didn't lap up the meager flames and transform their cherry-red seams into ladder fuel for my legs.

Even more, I wonder what our Lutheran and Baptist neighbors—mostly farmers from the Old Country who still spoke in thick German accents—thought of this little neighbor boy saying "Trick or treat" in a bright-red devil's costume. It's probably a good thing that they didn't know about my fireside dance.

10

Elk Camp

At the outbreak of a fire they would have seen
game fleeing. Later they would have basked in
the glow of the dying embers, and picked up
partly charred animals and fruit from the ashes,
savouring them. In this manner they would have
learned to appreciate the advantages of boiling
and roasting, which added not only to the taste
of meat but also, more importantly,
to its preservation.

—JOHAN GOUDSBLOM,
Fire and Civilization

It's still pitch-black as Robert and I stop at Sweetheart Donuts in Bend. A portly, bearded man behind the counter fills our thermoses with steaming black coffee and stacks pastries in a pink, tissue-lined box. I grimace as I notice the maple bars being slapped one on top of the other, the golden icing welding them together like chinking between logs.

In an odd mix of urban and backcountry, streetlights and headlights reveal other rigs outfitted like ours—tarp-covered camping gear strapped to bloated luggage racks or heaped on top of over-loaded trailers. With a little luck, we think to ourselves, they won't all be destined for our Southeast Oregon elk-hunting grounds. They aren't.

Robert, who also engages summertime fire, drives his parents' orange-and-white Jeep pickup, since my Datsun is far too puny for the formidable task ahead of us. This year we tow a trailer, having accumulated such a cache of supplies that we probably look like a Hudson's Bay Company entourage headed to some great fall ren-dezvous. We have 150 miles of sagebrush to traverse this day be-fore we arrive at the banks of Coffeepot Creek, where we'll set up our elk camp and hunker down for the next week, the fourth week of November.

October brought layoffs for all the BLM's seasonal work force. Those involved in prescribed burning worked the longest. A few folks transferred to the Forest Service to assist with their fall burning program. I opted for an early layoff. Now it's the end of November, and I've begun to settle into my winter mode of exis-tence—balanced among significant dates in the state's hunting regulations, meager attempts at running a trapline, and a backlog of unread books. Such are the preparations for winter. I resonate with the words of Søren Kierkegaard, writing under the pseudo-nym Johannes Climacus:

> I had been a student for ten years. Although never lazy, all my activity nevertheless was like a glittering inactivity, a kind of occupation for which I still have a strong predilection, and perhaps even a little talent. I read much, spent the rest of the day idling and thinking, or thinking and idling, but that was all it came to; the earliest sproutings of my productivity bare-ly sufficed for my daily use and were consumed in their first greening.[1]

Fall is a time for thinking and idling. At least it is the inception of such a time. But before I fully settle into that mode, there is elk season.

The slow retreat that fire began in late August has culminated in the virtual extinction of wildfire. Ever colder, wetter, and darker days and nights have all but annulled fire's ability to expand of its own accord. Ever more slash fuel or Alumagel is needed to get piles of brush to burn, and what underburns are set—to reduce fuel beneath the forest canopy—are allowed to creep around at night after burning crews retire for the evening. It's a strange sight to drive through the forest on a November night. Smoke hangs low as the cool night air sulks in sheltered ravines. Little glows twinkle on both sides of the road, like the warming fires from some mass encampment. Pitchy resins from some stumps burn exceptionally hot, even as temperatures dip well below freezing. And not a bit of yellow or green Nomex is in sight.

As with all seasons of the year, fire endures during fall, though in different forms and intensities. If nature is unable to froth up clouds into sparks of lightning, we'll supply our own ignitions.

For Robert and me, daybreak comes amidst an ocean of sage halfway between Bend and Burns. We realize we should have left earlier, for we'll have to stare into the sun for another seventy miles. Several months earlier we roamed this country in our yellow fire engines, looking to the south for cloud buildup, and then scanning the horizon for drift smoke. Some of the fires that began in the desert, when pressed by strong winds, made valiant runs toward the ponderosa-flanked Ochocos. Today, we just stare east, think about what pieces of camping gear we may have forgotten, and wonder if either of us remembered matches.

We drive past a historic marker in the Silver Creek Valley, near Riley, marking the 1878 skirmish between the U.S. Cavalry and the Bannocks and Paiutes, known as the Bannock War. I wonder what sort of fires burned the morning of that battle. As the morning inversion broke on that day, instead of the clean warming fires of sage and bitterbrush, smoke rolled from gun barrels. Now cattle sluggishly graze on each side of the road, unaware of all but the next clump of grass.

Another half hour and we're in the little mill town of Hines, population fourteen hundred. Hines spills over into the slightly larger town of Burns—the seat of Harney County—where we stop to refuel, decoffee, recoffee, and get back on the road again.

Fifteen miles later, we turn north off U.S. Highway 20 and begin paralleling Rattlesnake Creek. The headwaters of Rattlesnake Creek are in the extreme southern reaches of the Malheur National Forest. Rattlesnake Creek averages no wider than a shovel handle's length across, and maybe six to eight feet in the larger pools, yet it holds plump redband trout under what sheltering willows have prevailed against Herefords and Angus. Eventually, Rattlesnake trickles out into the Malheur Lake basin, a gigantic caldera, now a landlocked catch basin for streams draining the Malheur and Ochoco Forests to the north, and the fault-blocked Steens Mountain Range to the south.

The area is familiar. When I was twelve, I shot my first mule deer buck in the mountain mahogany and juniper fringe country just above Coffeepot Creek. At that time we had family connections with the owner of a ranch on the site of old Fort Harney, along the banks of Rattlesnake Creek. I haven't hunted deer there for some years, and the ranch has since changed hands. Fortunately, the new owner agreed to let us hunt elk on his land.

Before setting up camp, we continue past the private alfalfa pastures of Rattlesnake Creek, over the cattle guard marking the Malheur National Forest boundary, and up into dark timber below King Mountain. There, in a dog-hair thicket of pine, we cut firewood for the several points of combustion that will keep blood and coffee flowing through our veins in the days to come, and fry the elk liver and onions that we wistfully prepare for year after year.

Finally, our rig loaded with firewood, we backtrack south out of the forest, down into the sage and juniper draws of Fort Harney, and then northwest through a few barbed-wire gates into Coffeepot Creek Canyon.

Down in the bottomland, subdued grays and greens and yellows dominate. Leafless, gray willow branches flare out along the naked, cut banks of the creek. Spent, light-yellow leaves lie about on the ground and hang suspended in some of the branches: evidence that the seemingly lifeless *Salix* sticks aren't dead. It didn't seem as if the cut banks were always this high. Even if memory deepens reality, I still wonder to what extent overgrazing and consequent erosion have contributed to the willows being so high and dry, the water running so low and meager.

The ridges on our east and west jut to the north. Their southern tips trail off into sage and juniper plateaus and, like Rattlesnake Creek to our east, taper into the Malheur Lake basin, home of the Malheur National Wildlife Refuge. Parallel to camp, and high on the ridges, is the transition zone between juniper and pine—BLM land. There, amidst thick patches of mountain mahogany, the elk like to bed down. The beds are easy to find—large depressions of packed ground scattered between the junglelike mahogany and sprinkled with little piles of elk droppings, some black and fresh and sticky, others crumbling gray marbles. The ridges continue north into thickly timbered stands of ponderosa and fir—national forest land.

We put up our tent, a sea-green canvas monstrosity suspended from an arachnid-like frame of aluminum poles. To accommodate a woodstove, I had cut a hole in the ceiling, in which we secured a diamond-shaped metal plate through which the stovepipe would protrude. The previous year we brought a cabin tent that required wood poles to support its roof and walls, much more rustic (and more intuitive to set up) than this year's shelter.

Keeping warm is a necessity, particularly in light of forecasts predicting snow and subzero temperatures. We set up our beds in one end of the tent—foam pads and sleeping bags. The other end is for cooking and heating. Slightly to the right, along the rear wall, is where we place our little sheet-metal sheepherder's woodstove. Within its gut will throb a treasure trove of incandescent jewels that will thaw us from our predawn rigor mortis; it will also cook our meals, one being the Thanksgiving turkey that slumbers in our ice chest. This stove is the center of November fire along Coffeepot Creek. Any of the other fires that we build will radiate out from this flaming core of our elk-centered universe.

Each fire has its own ritual of starting and tending, but, by far, the woodstove requires the most preparation. Sometimes, in a pinch, crumpled newspaper is the tinder by which larger pieces of kindling are ignited. Usually, though, two split pine logs are laid parallel inside the dirt-lined belly of the stove, between which are nested curled slivers of pitchy pine or papery juniper bark. Over these, and between the split pine logs, we place still larger slices of pitchy wood. Finally, cradled above the tinder nest is cured juniper kindling. We have to leave the tent door open while kindling the first fire, for smoke sloughs off the black, newly painted stove surface like the diesel exhaust from a semi.

Mornings begin with the probing beam of a flashlight and a strike of a match to tinder. Generally, a single match is all that is needed, definitely all that is hoped for as one stands barefoot and clad in long johns in the icy morning air. With the stove's door closed, the newborn fire gasps its first breaths and is coaxed upward by a draft issuing through the door's vent hole, a breeze hellbent on climbing up the stovepipe. Once all the wood is ablaze, it's time to insert larger pieces of pine. If the fire is stoked sufficiently, really more than sufficiently, the stove's thin metal walls begin to warp. Waves of discoloration contour the metal surfaces, and, in the dim light, a red glow takes root on the stove's walls and grows with every additional piece of fuel. A signal to stop stoking.

Our central heating is not without casualties. As I've discovered, polypropylene long johns have a fairly low melting point—especially when in direct contact with hot metal. Once, in an attempt to thaw out our coffee, which had congealed overnight into a block of frozen Folgers, I slipped the entire pot into the sheepherder's oven, only to find it a bit later thawed, but with its plastic inner basket also well melted. This pot would thereafter be dubbed the Meltmaster.

Outside, off to one side of the tent is our general purpose fire ring. We scrape the ground clear of grass in a circle about the size of a garbage can lid, hoping this will avert any advance of the fire toward our tent. We use this fire for some cooking, a lot of coffee boiling, a good share of warming, and much gazing.

Our third fire is reserved for our cast-iron Dutch oven. Here we dig a hole a foot or so deep through the matted grass, and about two feet across. We stoke the fire with alder and mountain mahogany, and, after an hour, the bed of glowing coals resembles the throat of an active volcano. Then we lower the Dutch oven (replete with its savory contents of potatoes, carrots, onions, broth, and diced chunks of venison from last year's deer hunt) into the pit and cover it with the molten wood.

Our fourth fire is the trapper's fire. We've brought along a gunnysack full of steel traps with the intent of running a short trapline for coyotes. Before making our sets in the field, we must sterilize the traps to get rid of human scent. We've read about such things in books. So we place four rocks around the rim of our general purpose fire, upon which we set a metal five-gallon bucket filled with water. In the water we jam traps and a potpourri of sagebrush, rabbitbrush, and juniper branches. Eventually the soup simmers and boils, and, after a half hour, the traps are ready to be air dried and set in nearby ravines. Hudson's Bay revisited.

Nary a piece of hair is pinched by our traps this year. Regardless, in the mix of fire and traps, and a trapper's fire in the midst of elk country, there exists an inarticulate quest of young men feeling life and death in its cold and hot reality. We read Robert Service by lantern and drink Yukon Jack by the finger. But we really don't need to fantasize about being any place other than where we are. We're in a place that we know, yet that is seasonally new. We're in a place where, two months earlier, rutting elk scraped the bark off willows where our tent is now pitched, a place where coyotes howl and where one can die of hypothermia if unprepared.

While hunting, I like to carry enough gear in my Cordura rucksack that I could spend the night in the field, if I had to. This consists not only of food and water but also during late fall, a space blanket, flashlight, matches, flint, fire-ribbon, plastic baggie full of pitchy wood slivers, knife, and hatchet. Being injured is one reason that a person might have to hunker down for the evening; a downed elk is another. Robert and I have the understanding that if either of us gets an elk late in the day, or around dusk, we should probably just build a fire and spend the night at the kill. We'd dress and quarter the animal by ourselves and then hike out for assistance in the morning. Far from being dreaded, a night with the kill is anticipated with eagerness.

But on this trip, a survival fire is never kindled. Our trapper's fire is kindled only once. The Dutch-oven caldera is evoked a couple of times. The campfire is built when we have the energy to do so after a long day's hunt and it isn't too cold outside in the late-afternoon darkness. The sheepherder fire is built and nurtured daily. In all our fires, the act of tending allows us not only to appreciate aesthetic qualities and transforming processes of fire, but also to play a role in the initiation and sustenance of these events ourselves. This, too, like the watching, does not embrace mere utility but is valuable in the participation in the event process, the happening, of combustion itself, regardless of whether or not coffee boils or numbed fingers thaw by its heat. It's a good thing to have a hand in one's sustenance in a less mediated fashion than merely turning the dial on a thermostat. One needn't have a very sophisticated understanding of thermodynamics to appreciate the value in unlocking, and keeping open, wood's hidden window into the fiery potential of matter.

Winter

11

Settling

The fire confined to the fireplace was no doubt for man
the first object of reverie, the symbol of repose.
One can hardly conceive of a philosophy of repose that
would not include reverie before a flaming log fire.
Thus, in our opinion, to be deprived of a reverie
before a burning fire is to lose the first use and
the truly human use of fire.

—GASTON BACHELARD,
The Psychoanalysis of Fire

On the grassy slopes beneath still-golden wheat stubble, and in the lava highlands amidst stunted sage rooted between chunks of mossed-over basalt, free-ranging fire no longer burns. The chill of Arctic air snaking down from British Columbia has seeped into the high-desert flora. What fire now exists in the ever darker and cooler and damper days of winter is contracted; fire—whether the object of watching, tending, or engaging—now subsists only where humans kindle it. It may still burn intensely, but only when confined to a woodstove, campfire, fuel-cajoled slash pile, or flickering candle wick. Nature may still throw down an occasional lightning bolt in the Intermountain West, but the icy embrace of mid-December swallows up the electrical surges in virtual anonymity. Winter is a time for settling. It is a time to take stock, a time to survive, a time during which some will die. It is a time to think and a time to idle, a time to meditate.

There are parts of the country, and of the world, that see a resurgence of wildfire during my season of winter. For them, "winter" is probably not the appropriate title for such a season; as far as wildland fire is concerned, for them it is still summer. Some such areas may even be visited by more than one fire season in any given calendar year. But here, amidst bitterbrush and ponderosa, winter will pile up until it melts down, and only in the seasonal drought of six or seven months hence will summer really return.

In many homes in December, Christmas trees are adorned with multicolored bulbs in symbolic, annually renewed expectancy of the coming, and already having come, Savior; the bright-blue berries on the juniper also signal an advent of sorts—contemporaneous with the religious coming, testament to enduring color amidst expanses of white powder, and fodder for those who make wreaths for Christmastide. Ribs showing through the emaciated chests of mule deer and Rocky Mountain elk testify to winter's ability to draw heat out of not only fire, but also life. Accordingly, the inarticulate hopes of many an animal's life will be consumed not in fire, but in snow. The cold can be hard; it can also be endured; with a stack of dry wood and an adequate shelter, it can be savored.

Eventually, though, winter will pass, as it always does. When it does, spring will arrive with its own hopes and expectations. In the meantime, just as the Messiah may bring the fulfillment of hope even amidst further hopes for what that One might accomplish, winter carries hopes that will find fulfillment both in the season from which they arise, and in seasons beyond. And many such hopes and actualities are born of fire.

✦ ✦ ✦

Heat. On sweltering August days the blinding white disc suspended in the blue desert sky is a most cursed sight; in winter, the sun
still hangs in the sky, though lower, and its rays are now a coveted
treasure, but not nearly sufficient to keep naked-skinned hominids alive. So we build fires. Of course, modern technology affords
more than one way to create heat. There are baseboard heaters,
water-filled radiators, oil-filled radiators, electric and oil furnaces,
heat pumps, and natural-gas and coal stoves. But the allure and
romance of the wood flame as a most fitting object to warm up to
remains deeply entrenched in the human psyche. The reverie one
experiences while looking into a campfire reinforces the fact that
there are more reasons for staring into a fire than to warm appendages. Whether under open sky or under vaulted ceiling, the
desire to heat one's body before a real fire is of no minor significance.[1]

I despise fake fireplaces. A whole panoply of these devices
abound. Some mimic genuine combustion better than others. All
are inadequate to bear the values and evoke the reverie indicative
of naturally self-sustaining fire. How ridiculous are those counter-
feit log fires crafted out of brown plastic, replete with red translucent sides through which shine rotating lights ensconced in their
hollow cores. Such devices are often inserted into existing fireplaces whose owners hope to lower heating bills while maintaining a homey, romantic ambiance. With electric heaters discreetly
concealed, hot air blows out of adjoining vents to simulate the
heat produced by real fire. To the makers and purveyors of such
devices, the best of both worlds is achieved—a visual equivalent of
fire, and clean, hassle-free efficiency. At the most, only the latter is
achieved.

Other impostors, which burn natural gas, also fail to bear the
same awe-invoking value of naturally burning wood fires. Restaurants are notorious for such party tricks. Jets of flaming gas spray

out around loglike cement cylinders. They bespeak a tacit recognition that there is something desirable about the real thing, but wouldn't it be far more honest just to warm oneself by the gas flame minus the fake log? At least, then, the flame would be authentic. Better yet, instead of a log, have the flames emanate from a whitewashed, miniature propane tank—a true symbol of from whence the heat comes!

◆　◆　◆

Under the enveloping chill of winter, families gather in their homes around whatever heat source is available. Until only recently in history, the source of warmth was wood fire. It was a family endeavor to gather the wood and to keep the fire burning. And, besides the heating, such fires were necessary for nourishment—their open flames cooked food. In a very real sense, the fires of hearth and home carried and promoted many a family tree. It's no wonder that they still carry rich symbology of home and family and connotations of sturdy virtue. To maintain the fire takes more than a flick of a switch; it requires forethought, nurturing, and care. Lack of any one of these traits might lead to a cold pile of ash in the firebox, or a billowing flue that engulfs the entire abode in its creosote-raging blast.

To see the fire licking at the spit-skewered meat or the blackened cast-iron pot reminds one of where the heat originates and how the food is cooked. Under the calm of a January moon that pours soft blue light on whiskery tufts of bunchgrass, the orange coals of the hearth evoke meditation. Tagging along with each glance into the flaming juniper is the peripheral memory that this very fire prepared dinner, that similar fires sustained the lives of generations of one's forebears, or that this very fire now gives warmth to a scantily clad body that would die of hypothermia on the other side of the wall of the cabin or house. The grand scale of combustion in summer is condensed and confined, even embraced, within the intimacy of one's home.

What may once have stood as a symbol of moral ambiguity on a smoky July day in the Rocky Mountains is now much more transparent: Fire is necessary and good. To the extent one remembers that the yellow-and-orange mass spiraling and twisting on the hearth originated from life, the reverie will include an awareness that fire is life, and that what was once alive is now afire.[2] Looked at metaphysically, there may be the vague consciousness that the flames are unlocking a substratum of reality (or an intimation of a reality) unknown most of the time: If there is fire in the wood, what else might there be of which we are unaware? However, I suspect that most of the time, as one dips into a trancelike state before the flames, such thoughts rarely come to analytic fruition. Instead, the experience is of a much more mystical sort, intuited, felt, known in one's bones. Magnetic.

It is a symbiotic event. The fire only exists because someone started it. That someone who kindled the fire lives only because the fire now feeds on itself and warms the frame and cooks the food of its human tender. The fire does not care—consuming whatever it can—but its lack of consciousness makes the relationship no less symbiotic. If there be a God in heaven, or an omnipresent God not altogether absent from the wood, then the very mystery that this symbiosis hinges on finds its ultimate repose in religion; seeing the flaming wood at least invites such wonders.

And, since snow blankets the particularities of the landscape, concealing the diversity that made fall a season of reflection, winter turns reflection inward, nurturing a season of meditation.

✦ ✦ ✦

As I've suggested elsewhere, fire is a powerful symbol. Such symbolization need not be limited to the season of winter. However, to the extent that all life is turned inward during winter—physically and spiritually—this season breeds a propensity to weight fire with more expressive value than it might otherwise hold.

For both philosophy and religion, the point of slender light jutting up from a candle wick is a symbol of illumination. The light that it provides, standing on a darkly oiled desktop, or on the silver or gold fixtures of an altar, draws attention to the illumining side of truth, love, hope, or events that the wax itself might symbolize. Whether in a menorah or a circle of candles embracing the weeks of Advent, the lighting of each candle actualizes in a burst of heat and light—palpably and visibly—some event or reality beyond the flame itself. The darkness of a room heightens the concentration upon the flame, which illumines this reality. The symbols conceived in this season will move through the year—retaining their character while pointing to the future—only to once again find their renewal and intensification in dim sanctuaries and study nooks, or beside living-room hearths three seasons hence.

The heat of the flames also symbolizes renewal. Not only will heat again flow out across the frosted landscape as days lengthen and months wear on, but this renewal speaks to other sorts of moral renewal. Oftentimes before renewal must come purification, and fire is the great purifier. This is to be taken very literally. Like renewal, this ability to purify makes fire an unsurpassed symbol, and metaphor, for all sorts of purging, cleansing, and purification, and in the most radical sense. This is nothing new. Didn't the voice of the one who cried in the wilderness of Judea speak of chaff being burnt by "unquenchable fire"?[3]

Examples could be multiplied of how fire has, through history, taken on symbolic meaning. Yet I suspect that all such significations carry with them the mundane—though truly extraordinary —realization that fire itself is a mystery. It reveals, yet its silent brightness conceals how it sustains itself. Thus each moment of apprehending the contracted flame of symbolic fire forms a spiral of meaning. Each moment carries within itself the vestigial memory of the directly apprehended significance of fire, and each present moment of encountering the flame—a happening—carries with it all those symbolic meanings that heat and light evoke or awaken.

◆ ◆ ◆

It was strange to see fire blasting out of the green sod of the White House Ellipse. But Washington, D.C.—as political pundits are well aware—is a town rife with irony.

Earlier in the day, my wife and my mom and I toured the president's house to see the Christmas trimmings and partake of the festivities of the season, enshrined at a national level.

In the evening we returned to the White House grounds to see the fully illuminated national Christmas tree. We fought our way through the crowds onto a little path encircling the gaudily lit tree. Along the path were smaller decorated trees, each representing a state, and each decked out with ornaments crafted by that state's residents. Further on, there was a manger scene, a half-dozen beleaguered reindeer, and what appeared to be a bonfire submerged in the lawn, large enough to barbecue an entire cow.

A rail fence bordered one side of the ten-by-six-foot brick-walled, rectangular grave of flame and coals. Behind the neck-deep pit were a pile of thick logs and a front-loading tractor that was apparently used for stoking the fire. Truly presidential. An interpretive sign told us that this was *the* Yule log. For many of the people who took a merry glance at the holiday flame and continued on, the seething pit was little more than a quaint symbol, albeit one to keep scampering children from glissading into.

I stared for a long time into the glowing mass that burned in the lawn. It was a vestige of wildness, even hemmed in by brick and

wooden borders. It flared under the sky—wild black sky—silhouetted before a horizon that bore up hard white monuments and
domes illumined in the night.

Our national Yule log was not quite the same as the fires of Yule
that Scandinavians long ago kindled in celebration of the winter
solstice. Nor was it exactly like the fires that Christians built as
symbols of the Son coming and already having come to earth.[4] In
fact, most people who walked by the rail fence were probably
quite unaware of the glowing lineage of that log, burning where
Union troops once camped. Nevertheless, the Yule log is a symbol
that people expect, like other symbols of the so-called holiday season—symbols that have lost the richness of their signifiers to
generic associations with Christmastide. And flaming up there in
the Ellipse, next to the national Christmas tree, the Yule log was
imbued with national significance: It was the seasonal fire of a na-
tion. It was warmth and light, something to thaw cold fingers next
to, homey, a more unrefined cousin of the lights that glowed in
civilized, multicolored arrays on the little and large Christmas
trees. Maybe that's why it was on the periphery.

12

Bracketed by Fire

Heating with one's own wood may be no more
"authentic" than central heating, but it offers a
far clearer metaphor. Heating with wood is very
much a participatory activity. In the yearlong
cycle, from flagging trees for culling to the rich
glow of oak cinders of a winter's night,
the subject is constantly present and nature is
directly present to him, both in the hardness
and in the caressing softness of its reality.

—ERAZIM KOHÁK,
The Embers and the Stars

Winter lays bare the complexities of fire. What is magnified
during summer on the scale of acres and sections is contracted to
the size of square inches and feet during winter. In the stark simplicity of light against leafless cold, individual characteristics of
fire, and how they affect our lives, are often more obvious. Stalks
of flame grow out of *that* piece of wood; a draft increases if I pull
on *this* lever; unless I feed the fire, and do so with care, it will die,
from either starvation or suffocation.

The seasons of fire are a bit like aesthetic judgments of natural beauty—certain qualities of nature, and fire, appeal more to some individuals than to others, and it's not always clear why. Some people despise the aridity of the high desert landscape. They say it's lifeless, boring, dry—indeed, worthless. For them, rain forest and meadow and wildly extravagant flower gardens exhibit true beauty. Of course, any such judgments hark back to the age-old question of whether there is some objective, universal standard of beauty, or whether beauty lies purely in the eye of the beholder. As near as I can tell, the truth dwells somewhere in between.

To press the issue further, one's aesthetic judgments, and soul, often mirror qualities of the physical landscape. Some people love busyness; they're attracted to landscapes rich in verdant particularities. Others feel overwhelmed, or at least hemmed in, by what they perceive as clutter; they seek out austere places with a view. Both preferences are sometimes blind to instances of complexity and simplicity in all landscapes. While many of the drier regions of the Intermountain West do lack the scale of luxuriant plant life found in Pacific Northwest rain forests, a wide variety of flora still call the Great Basin and high desert home. From miniature lichens and wildflowers to rodents and insects that scurry around in the duff beneath papery-dry sagebrush trunks, there is life, and life in abundance. But because there is ostensibly less clutter in the desert, if you really want to, you can focus your attention on the particularities. And this is the secret of wintertime fire.

Similarly, the lives of some people have an affinity to the fire of one season more than to another, even if we're all touched by fire in all seasons to varying degrees. Whether or not one is *attracted* to the fire of any particular season, that the fire of that season has, in fact, framed the lives of many is undeniable. And the lives of many have been framed by the fires of winter.

✦　✦　✦

Though the Sitka spruce forests of the Alaska Panhandle are markedly different from the tundra expanses of the vast interior, they can be no less forgiving to the unprepared in winter. In an old typewritten manuscript, my grandfather Ludwig Berg wrote of a duck hunt that he and two others took along southeastern Alaska's Inside Passage. It was 26 December 1909 when the three set out in a sixteen-foot skiff from Wrangell, en route to Berg Bay, twenty miles southeast of town and along the mainland. At Berg Bay, my grandfather and great-grandfather owned a cabin that they stayed in while prospecting for gold and silver in the nearby Aaron Creek watershed.

The morning after they arrived at the bay, the trio of would-be hunters discovered the bay and surrounding sloughs covered with a thin skin of ice, and no birds. "We did not worry," wrote Granddad, "because we had enough provisions along for seven days, and it may turn warmer in a day or so. To our surprise it turned colder and the temperature dropped below zero the next day."

On 17 January—three weeks into their ordeal—the group decided to press ten miles inland, along a trail that weaved through tangles of devil's club and swampy muskeg, to their upper mining cabin. There they hoped to find provisions left over from their summer prospecting season.

We finally made upper camp at 6 P.M., and were happy to have made it. On arriving at the cabin we found the door open and snow blown inside. Looking around we found that nearly all the wood we had cut for storage for next season's operation was used by some late fall trappers, catching marten. Then we found that they had left the stove pipe on the roof, and it had blown off. We had a time finding the pipes, but we finally saw part of them sticking out of the snow. We finally raked the snow off the roofplate and set up the pipes and, finally, had a fire made. It was then dark. We then shoveled out the snow and closed the door, and believe me we were tired, but the fire felt so good that we were glad to be alive.

Their days revolved around eating and keeping warm. But even in the death grip of ice, it's still possible to get too warm, as one paragraph from the story makes clear:

> About noon the 19th, the little snow left on the roof had blown away, and we heard some cracking on the roof. We ran outside and found a small blaze between the shakes. Luckily having some snow thawed on the stove, this helped us to put out the blaze. Had this been at night or when we were away, then we would have lost our cabin, and we would not have been able to survive in the zero weather without a shelter.

During winter, such things are possible. On 28 January, thirty-four days into their ordeal, the three flagged down a boat anchored just beyond the frozen bay and caught a ride back to Wrangell.

Who knows how much of Granddad's story is hyperbole, but that their cabin did combust outside the firebox and that the consumption of one's shelter in the middle of an Alaskan winter is a serious affair are reasonable enough assumptions. Even in winter, fire remains a tentative ally. By conspiring with cold, fire's once comforting heat can—and often does—kill.

◆　◆　◆

During the winter, while I was growing up on the Heinz place, my family burned Presto logs in the trash burner of the kitchen stove. Because the firebox of the stove was too small to fit the yellowish logs in their entirety, I was given the privilege of chopping them into stove-palatable pieces.

A door in the kitchen opened to a rough, wooden-floored storage room, beyond which was another small room that served as the kennel for our two black Labs. The musty storage room was unheated, and there we kept our garden tools, wood, and Presto logs. We bought Presto logs at a grocery store in Molalla, and after hauling them home in the back of our 1972 Ford pickup, I would unload and stack them in the back room. There was a large round of well-tenderized wood on the right side of the room that served as a chopping block. In the evenings, following the hasty completion of my homework, I would retire to that chilly little room to slice up the logs. Many parents are afraid to allow their children to wield sharp instruments; I was given a leather-handled Estwing hatchet to entertain my eight-year-old curiosity. So on winter evenings I went out back, lay Presto log after Presto log on the chopping block, sliced them into two-inch-wide wafers, then carried them in a cardboard box to the stove for burning.

Slicing up those stubby golden rounds taught me that work can be fun. There was a sense of accomplishment in seeing the neatly stacked rows of whole logs evaporate as the nearby heap of crumbly disks grew to a third grader's waist level. Most of the time supply exceeded demand, as I chopped away oblivious to how quickly the butterlike pads were melting in the stove. I enjoyed engaging in an activity that was not merely child's play; the logs that I split helped keep my family warm.

I felt like one of the pioneers that I had read about in my grade school primers, pioneers for whom chopping wood and living off the land were the only options. I especially loved stories about the Pilgrims and always longed to hunt turkeys just as those New Englanders did, toting brightly colored toms and black-barreled blunderbusses back to homes with thatched roofs. Much to my chagrin, there were no turkeys in our brush rows and woodlots, nor did I own a blunderbuss; however, I did own several cap guns, and the local ring-necked pheasants looked sort of like turkeys. But besides wild game, firearms, and era-appropriate garb, I knew that I needed to partake in the labors of fire to truly be like a pioneer. Even though Presto logs were hardly the fuel that nurtured the fires of either European immigrants or those natives who netted salmon and stalked deer long before Captain Robert Gray penetrated the mouth of the Columbia, they were similar. They burned and they required a sharp ax to ready them for burning. A sharp ax I had.

◆ ◆ ◆

I suppose that anyone might be able to find examples of ancestors whose lives have been punctuated by fiery lore. And I suppose that those who find their employment and sustenance at the edge of fires throughout the seasons of the year may be more clued in to such punctuations.

I recently discovered that one of my great-grandfathers, Louis Byers, who was allegedly killed by a mountain lion early in the late 1800s, actually died while serving time in the old Montana State Prison in Deer Lodge, Montana. This is interesting family history in and of itself, but, ironically, his denouement under a cold Montana sky was a function of the fires of winter.

In 1894, Louis operated a butcher shop in uptown Butte, where the Silver Dollar Saloon now sits. Other than that, I know little about his life, which probably isn't that uncommon for histories colored in shame, especially when the survivors are a wife and small children in a small town at the turn of the century. The source of shame became clear when a distant cousin of mine stumbled on the following paragraph in the 1 January 1899 edition of the *Butte Miner*:

Louis Byers, sent to the penitentiary here in 1895 for cattle stealing near Silver Bow, was killed yesterday afternoon while hauling wood to the penitentiary. While coming down a steep hill the wagon upset and crushed one of his legs beneath the knee and he died soon after from the shock. An inquest was held today and no blame could be laid to the contractors, as Byers was doing the work pertaining to the penitentiary.

Whether or not the wagon was assisted in its deadly course by the owners of the rustled cattle is worth pondering. Whatever the case, the ritual and necessity of keeping warm was, and still is, not without its hazards.

◆ ◆ ◆

Wood and fire both have a way of marking a youth for life. When my dad was in the eighth grade, living at Twelfth and Sandy in northeast Portland, he accidentally chopped off a couple of his fingers with an ax. One day while home for lunch, he decided to chop some wood in the basement. Later that afternoon he was to fill the lead role in a school production of *Huckleberry Finn*. (I'm not quite sure why he was chopping wood rather than practicing his lines, but who's to question the integrity of an eighth grader.) As he lifted the ax, it struck an overhead joist, which deflected it back down on his left hand. I remember Dad telling me the grizzly story of how, as his fingers bled profusely, he had to dial the telephone for an ambulance, since his mother and siblings were too shocked by the horrific sight of the rough-cut stumps to make the call for him. A doctor sewed the digits back in place. Unfortunately, Dad had to return to the doctor several days later, as gangrene had set in, in his left index finger. The doctor snipped the blackened finger off with a pair of scissors, leaving a stub below the first knuckle. Luckily, the repair to the other finger proved more successful.

World War II came. Dad tried to enlist in the navy but was at first turned down because of his missing finger. When the armed services became desperate enough to accept individuals with missing digits, Dad successfully signed on, although, according to Dad, because he was also missing a few teeth, he had to settle for life aboard a PT boat rather than the submarine duty he had hoped for. As a result, he roamed the South Pacific aboard *Elcos* and *Higgins*, spewing their own form of smoke from their massive Packard engines.

Thirty-five years later, when Dad retired from sheet-metal work, my folks moved to a sagebrush and crested-wheat plateau in Central Oregon, where Dad spent many a summertime afternoon in the backyard splitting juniper into kindling. He would set chunks of gnarled juniper on top of a larger pine round that served as a chopping block, whose chewed-up surface testified to its many intimate encounters with axes and malls.

On one of these days that Dad was splitting wood, a piece of juniper sheared off the block, toppling over onto his foot. For most people, this would be no big deal: a cut toe, a black-and-blue nail that might fall off but eventually would grow back. Dad's diabetes was not nearly so forgiving. The poor circulation in his leg refused to allow his toe to heal. It had to be amputated. Again, the incised region became infected, so the only recourse was to make the next cut where there would be some hope of healing, which meant just below the knee. Blood clots and strokes followed, and his health continued to deteriorate.

Trips to the hospital for more tests became frequent. I'm not sure whether the tests produced more accurate diagnoses or merely tested Dad's failing endurance. With each battery, he came home weaker than when he left. The tracheoscopies were the hardest on him. The final diagnosis was advanced lung and bone cancer. It got to the point that Mom couldn't take care of Dad at home any longer, and my being out fighting fire didn't help. So one day an ambulance came and took Dad to the hospital in Redmond.

With each passing day, my dad's strength seemed to evaporate, degree by degree. Breathing became ever more labored, his pale blue—now bloodshot—eyes a little more distant. One day, Dr. Detwiler walked into Dad's hospital room while my mom and my brother and I stood around the stainless-steel bed rail. "How are you doing, Ed?" he asked Dad in a slight Southern drawl. Dad had a hard time speaking, let alone speaking with a plastic oxygen mask on his face. The doctor reached down and lifted the mask from its perch atop Dad's salt-and-pepper-stubbled cheeks. "Let's get this damn thing off," Dr. Detwiler said. Which said to me, there is a time when the dignity of one's life, at the end of life, means more than plastic tubes and masks and metered gasses. The lifting off of that "damn thing" meant a lot to me, and it may have to Dad.

The doctor wanted to have a family conference to discuss Dad's condition. These meetings are the sort I had always associated with bad made-for-TV movies; they are also the type that some family holds each day. We found a quiet alcove in one of the hospital corridors where we could discuss Dad's options. The doctor mentioned that they could possibly perform surgery, but given Dad's poor response to the tracheoscopy, the positive results would probably be few and transient. The bottom line was that Dad did not have much longer to live. Without surgery, his options were either to remain in the hospital or to return home for as long as we could care for him. I had already decided that if Dad came home, I would take time off from work to help Mom care for him. Dr. Detwiler mentioned the possibility of hospice care, but he was keenly aware of the difficulty—some call it the "burden"—of taking care of a terminally ill person at home.

We decided to ask Dad what he wanted. Back along his bedside, we asked him if he would like to stay in the hospital or go home. In a high-pitched, barely audible voice, Dad pressed out— "Home."

Several days later, we would take him home, to the living room with the big picture windows that frame the silhouettes of Mount Jefferson, Three-Fingered Jack, and Black Butte.

Whenever the phone rings late at night or early in the morning, I fear the worst. Oftentimes callers have simply dialed the wrong number. Not so on a dark Wednesday morning of July 1989. It didn't take long for the ringing to stop. Minutes later, Mom knocked on the door of my bedroom, the bedroom of my high school years, the bedroom in which my dad wept when I went off to college, where I now took up residence when I wasn't at my guard station or fighting fire.

"It was the hospital . . . your dad just died."

It was really no surprise, but it was. Death always is. A 26 July entry in my Smokey calendar book says all that needed to be said, all that I could say at the time: "Dad died." Finality.

Mom and I didn't say much to each other as we drove the ten miles into Redmond. All details beyond the dashed yellow center line and solid white fog line were a blur.

The dim hallway through the patient wing of the Central Oregon District Hospital was peaceful in the predawn hours. The only lights burning were those over the nurses' station, a handful of quivering fluorescent panels that made a checkerboard of the ceiling in the hallway, the green EXIT signs, and a few weak bulbs above the head ends of some patients' beds. A nurse escorted my mom and me into Dad's room.

He looked like he was just sleeping, like he did twelve hours earlier following a large dose of morphine. I reached out and felt the soft, slick, fading-pink skin of his hand. It was warm. Though his heart lay still in his chest, his life—his heat—desperately clung to its home.

Later that morning, one of the first persons I called was the Prineville District fire management officer. I was unsure what to say but thought I should let someone know that I wouldn't be in to work for a few days. Someone on the other end of the line answered, "Hello."

"Yeah, Steve?"

"Yes."

"This is Dave . . ."

I was a little shocked and embarrassed that I could hardly speak. In a flood of tears and gasps, I croaked out that my dad had just died, and I needed some time off. Steve told me to take off as much time as I needed, which I did, even though it was fire season.

One of Dad's favorite poems was "The Cremation of Sam McGee," by Robert Service. It was only fitting, then, that Dad wanted to be cremated. So the following summer my family spread his ashes on a windswept promontory, just north of Mount Jefferson. One day in the future—maybe ten years from now, maybe ten thousand years from now—fresh volcanic ash will cover the spot anew.

Dad's cancer probably developed independently from the ailments splintering off that midsummer chunk of juniper. Years of cigarette smoking surely played a role. Handling asbestos as a sheet-metal worker also may have contributed. That my dad's deteriorating health was assisted by both these causes, one a source of combustion and one a substance to prevent combustion, cannot be denied. The wood, though, got the last word. It's strange how a piece of wood—prompted by preparations for wintertime chill—can bracket either end of a person's life. Maybe framing my dad's life in terms of fire is like envisaging herds of ungulates migrating in the cloud buildup of spring; maybe with another lens, or from another vantage point, some other set of variables could be viewed as defining critical points in his life. Then again, maybe there is something more literal and substantial about viewing his life—and the lives of others—by the light of flames. It might be a very profound truth to find ways in which something as ordinary as a stray ax blade or a falling stick of juniper can influence what we do and how we die.

EPILOGUE

The ordering, the same for all,
no god nor man has made,
but it ever was and is and will be:
fire everlasting, kindled in measures
and in measures going out.

—HERACLITUS, Fragment 37

For the lightning is ynough to shewe us
the glorie that is in him.

—JOHN CALVIN, *Sermons on Job*

The flames of midsummer have given way to the drifting snow of winter; the free-burning fires scaling forest and range in America's cordilleras have yielded to the kindled, hemmed-in fires of elk camps and ski-tour huts. Glistening white snow spirals down the rounded slopes of the Columbia River breaks, and protruding from beneath what had been acres upon acres of drooping, matted grass is a vast emerald carpet. The temporary metal stakes that marked the penultimate resting places of a handful of young firefighters have now been replaced by granite crosses, yet these, like their predecessors, will be overshadowed by oaks that continue to make a name for themselves as they reach toward heaven.

From the latent fire in water, lightning crackles through the sky and brings fire to the earth. This event process, this happening, will continue to be a literal and metaphorical partner of humanity into the foreseeable future. And, if Gaston Bachelard is at all correct that "to be deprived of a reverie before a burning fire is to lose the first use and the truly human use of fire," then I'll add: To lose fire is to cease to be truly human.

The rhythm of Qoheleth—speaking to all phases of life—applies equally well to the human encounter with fire, for there is

> A time to burn, and a time to cool;
> A time to converge, and a time to disperse;
> A time to kindle, and a time to extinguish;
> A time to embrace, and a time to combat;
> A time for ashes, and a time to sprout out of ashes;
> A time to celebrate, and a time to mourn;
> A time to build up, and a time to settle;
> A time to *watch*,
> > A time to *tend*,
> > > A time to *engage*.

These are the seasons of fire, and the seasons of humanity in a fiery world. They are also the seasons of earth under the starry heavens, under the sun whose warm rays grow plants—plants whose frames, once afire, wink back at the sky.

NOTES

PROLOGUE

The epigraph is from William Paley, "On the Elements," *Methodist Magazine* 2 (August 1819): 292.

1. Stephen J. Pyne, *Fire in America: A Cultural History of Wildland and Rural Fire* (Princeton: Princeton University Press, 1988), 530, also refers to fire as an event: "Fire is an event, not an element. It exists within a fire environment, without which it would perish."

2. Though there is a great deal of dispute over how long ago hominids began using fire, Johan Goudsblom, in *Fire and Civilization* (London: Penguin Press, 1992), 17, notes "that sufficiently solid evidence has been found in various parts of Europe and Asia to conclude that *Homo erectus* was using fire at least some 400,000 years ago—that is, long before the appearance of *Homo sapiens*."

1. BUILDUP

The epigraph is from Heraclitus Fragment 38, in Charles H. Kahn, *The Art and Thought of Heraclitus: An Edition of the Fragments with Translation and Commentary* (Cambridge: Cambridge University Press, 1979), 47.

1. Here and throughout, I use "moral" in a broad sense to include not only what is narrowly ethical, right, or just, but, more importantly, what is good. This conception of "moral" transcends mere obligation; it entails human flourishing, the flourishing of the nonhuman world, and all those factors that contribute to that flourishing and well-being. Similarly, see Charles Taylor's treatment of the good in *Sources of the Self: The Making of the Modern Identity* (Cambridge: Harvard University Press, 1989).

2. AXIS MUNDI

The epigraph is from Edward Abbey, *Desert Solitaire: A Season in the Wilderness* (New York: Simon and Schuster, 1968), 12.

3. CONVERGING

The epigraph is from Henry David Thoreau, *Walden* (London: J. M. Dent and Sons, 1962), 81.

1. See Pyne, *Fire in America*, 282, for an amplified discussion of the "10 A.M. policy."

4. ASSIGNED

The quotation is from Antonio Vivaldi, "Summer," *The Four Seasons*, in H. C. Robbins Langdon, *Vivaldi: Voice of the Baroque* (New York: Thames and Hudson, 1993), 61–62.

1. Daniel M. French, in *An Illustrated History of Central Oregon* (Spokane: Western Historical Publishing, 1905), 443–451, lists the following populations of turn-of-the-century Sherman County towns: Wasco, 700; Moro, 800; Grass Valley, 450–500; and Kent, 250. Today, the population of each of these towns is less than half of French's 1905 figures.

5. ENGAGING

The epigraph is from Gaston Bachelard, *The Psychoanalysis of Fire* (Boston: Beacon Press, 1964), 55.

1. J. G. Frazer, *Myths of the Origin of Fire* (London: Macmillan, 1930), 8–9, quoted in Gaston Bachelard, *The Psychoanalysis of Fire* (Boston: Beacon Press, 1964).

2. Using fire to speak of sex, and sex to speak of fire, might have been and undoubtedly is today a very profound crafting of metaphor. Fire—warming a hearth, dancing beneath a spit under the enveloping darkness of a night sky, or crowning in effulgent radiance through a dense forest—is an intense power that serves as a useful metaphor in any number of circumstances, especially those related to sex. Maybe we are more naive than the primitives whose psychological depths we attempt to probe

when we disallow them the privilege of speaking metaphorically. Or maybe they were agnostic about whether the flaming stuff under the sizzling spit was a constituent of sex. Even if they thought it was, this does not diminish the hypothesis that they viewed fire as multivalent. I am indebted to Nicholas Wolterstorff for my rethinking of the poetic/symbolic consciousness of ancient peoples.

3. In much of his scholarly work, Bachelard is engaged in what he calls a "psychoanalysis of objective knowledge." This entails "finding how unconscious values affect the very basis of empirical and scientific knowledge." Moreover, in *The Psychoanalysis of Fire*, 10, Bachelard writes: "We must then show the mutual light which objective and social knowledge constantly sheds on subjective and personal knowledge, and vice versa. We must show in the scientific experiment traces of the experience of the child. Thus we shall be justified in speaking of an unconscious of the scientific mind—of the heterogeneous nature of certain concepts, and we shall see converging, in our study of any particular phenomenon, convictions that have been formed in the most varied fields." While there is much to commend in such an investigation, the degree to which such a "psychoanalysis" is capable of getting to a core of objective, scientific truth is not altogether clear.

4. See Bachelard, *The Psychoanalysis of Fire*, 21–41, for his complete description of the Novalis Complex.

5. Goudsblom, *Fire and Civilization*, 19.

6. In an evolutionary sort of way (analogous to zoology's biogenetic law), it's as if "pyro-ontogeny" recapitulates "pyro-phylogeny." By this I'm suggesting that the child's learning about fire has an analogous precursor (and progression) in what Goudsblom calls the "original domestication of fire." To put it another way, in the life of any given human, that individual recapitulates the historical process of how humans first came to capture the flame. A child encounters fire first by watching it, then by feeling its heat, and only later through learning how to handle, kindle, and actively engage it—similar to the epic process of fire's original domestication, which would have been spread out over many, many years. For some human beings, this process of pyro-ontogeny might collapse into virtual simultaneity, and, admittedly, it holds better for the lives of some individuals than others. Regardless, it is a useful way to envisage the reacquainting of fire and humanity in each successive gen-

eration. See Goudsblom, *Fire and Civilization*, 1–12, for his discussion of the historical progression from passive to increasingly more active uses of fire, and as a contrast, Bachelard, *The Psychoanalysis of Fire*, 10–11, regarding the "social reality" of fire.

7. Regarding hearth fires, see, for instance, Margaret Hindle Hazen and Robert M. Hazen, *Keepers of the Flame: The Role of Fire in American Culture, 1775–1925* (Princeton: Princeton University Press, 1992), 217–218:

> The emotional attachment to flickering flames was particularly strong among people of English descent during colonial days, a fact that Benjamin Franklin took into consideration when he developed his stove with the open front. This penchant for fire watching survived the American Revolution with ease, and thereafter it flourished almost to the point of becoming a national obsession. Motivated variously by nostalgia for the good old days, by the belief that an open hearth represented family values and social stability, and by deep affection for the visual effect of fire, people of many backgrounds refused to allow open fires to disappear entirely from their lives.

8. My illustration is based on G. B. Shaw's reply to H. G. Wells in the *Sunday Express* (August 1927), quoted in John O'Neill, *Ecology, Policy, and Politics: Human Well-Being and the Natural World* (New York: Routledge, 1993), 158. There, Shaw describes an individual gripped by an amoral passion for knowing and seeing that which is new: "The vivisector-scoundrel has no limits. . . . No matter how much he knows there is always, as Newton confessed, an infinitude of things still unknown, many of them still discoverable by experiment. When he has discovered what boiled baby tastes like, and what effect it has on the digestion, he has still to ascertain the gustatory and metabolic peculiarities of roast baby and fried baby, with, in each case, the exact age at which the baby should, to produce such and such results, be boiled, roast, fried, or fricasseed."

9. Bachelard, *The Psychoanalysis of Fire*, 12.

10. Goudsblom, *Fire and Civilization*, 1.

11. Described in a witness statement by Bradley Jan Haugh, U.S. Forest Service and Bureau of Land Management, *Report of the South Canyon Fire Investigation Team*, 17 August 1994, A5–45.

12. Regarding the practice of setting fire for sport, and for the visual spectacle that it creates, see Pyne, *Fire in America*, 163, 337—The Pacific Northwest; 412—Southern California; 500, 502—Alaska. One particularly blatant example of using fire to titillate the human desire to see new things was the tradition of the Yosemite Firefall. According to a 1941 brochure, *The Firefall—Explanation and History* (Yosemite National Park, Calif.: Yosemite Park and Curry, 1941), this tradition was started by James McCauley in 1871 or 1872. It consisted of building a bonfire on the edge of Glacier Point, then shoving the red-hot mass over the edge to create a pseudowaterfall of incandescent embers and sparks. McCauley's son described his dad's lurid ritual:

> Just after the fire was kindled, he would take a gunny sack, saturate it with kerosene, put it across the end of the rake, light it, and wave it back and forth three or four times, then throw the flaming sack over the cliff, which would look like a ball of fire slowly dropping down the side. Three separate sacks were thrown over the cliff while the fire was burning, and three more separate sacks were lit and thrown over the cliff after the firefall had been shoved off.

Apparently the ritual caught on with park visitors. With great pride, the brochure goes on to say:

> At nine o'clock each evening during the summer season the stream of glowing embers pours from the edge of Glacier Point, 3,254 feet above Camp Curry. This sight of surpassing beauty is visible from most open areas in the eastern end of the Valley, and long before it occurs hundreds of visitors have sought vantage-points from which to observe it. Sometimes the embers pour straight down the cliff, gradually spreading fanwise as they approach the ledge below; again the stream of fire waves back and forth in its descent in the manner of a windblown waterfall. It continues for several minutes, and as the last mass of embers is pushed over the cliff, a shower of sparks and flame arises momentarily. Thereafter the glow fades very gradually and dies away, leaving the cliff to darkness and many a watcher close to tears.

13. I was first introduced to the phrase "dancing with fire" by smoke-jumper Murry Taylor during Winter Fishtrap 5, "Fire," a writing conference held at Wallowa Lake, Oregon, 24 February 1996.

6. NEW SHOOTS

The epigraph is from Norman Maclean, *Young Men and Fire* (Chicago: University of Chicago Press, 1992), 300.

7. MAN AGAINST FIRE

The epigraph quotes Doug Maxwell, narrating his experience as fire boss of the Canyon Creek Fire, in the film *Man against Fire* (USDA, n.d.).

1. See, for instance, Pyne, *Fire in America*, 237ff., and Michael Thoele, *Fireline: Summer Battles of the West* (Golden, Colo.: Fulcrum Publishing, 1995), 19–33.

2. William James, "The Moral Equivalent of War," in *The Writings of William James: A Comprehensive Edition*, ed. John J. McDermott (Chicago: University of Chicago Press, 1977), 662, 664. Echoes of Theodore Roosevelt stir in the background of James's remarks. See Theodore Roosevelt's speech "On the Strenuous Life," as delivered to Chicago's Hamilton Club on 10 April 1899, in *A Treasury of the World's Great Speeches*, ed. Houston Peterson (Chicago: Spencer Press, 1954), 655–660. In that speech, Roosevelt advocated the need for Americans to embrace lives of "strenuous endeavor" marked by "manly and adventurous qualities." Failure, on either a personal or national level, is a sign of weakness and unmanliness; hence, he enjoined Americans to *confront* problems and conflict (some of the most pressing during his time being America's political "responsibility" in Cuba and the Philippines). Roosevelt didn't limit the virtue-making realm of strife—or even strife, per se—to the military, for the strife must be justified. Rather, the strenuous life is embodied in a readiness and willingness to risk and sweat and toil and even die for the sake of duty, national honor, and righteousness.

3. James, "The Moral Equivalent of War," 663, 668–669.

4. A standard work in this area is Joseph Campbell's *The Hero with a Thousand Faces* (Princeton: Princeton University Press, 1973).

5. Jan Mejer, "Wildland Fire as a Liminal Environment," paper presented at a conference of the International Association of Wildland Fire, Coeur d'Alene, Idaho, November 1995.

8. DISPERSING

The epigraph is from the National Wildfire Coordinating Group, *Fire-line Handbook* (Washington, D.C.: NWCG, 1989).

1. See Pyne, *Fire in America*, 71–83, for a good discussion of the extent to which American Indian use of fire shaped landscapes and biota in North America. Also see Stephen W. Barrett and Stephen F. Arno, "Indian Fires as an Ecological Influence in the Northern Rockies," *Journal of Forestry* 80 (October 1982): 647–651; Thomas C. Blackburn and Kat Anderson, eds., *Before the Wilderness: Environmental Management by Native Californians* (Menlo Park, Calif.: Ballena Press, 1993); Robert Boyd, ed., *Indians, Fire, and the Land in the Pacific Northwest* (Corvallis: Oregon State University Press, 1999); George E. Gruell, "Indian Fires in the Interior West: A Widespread Influence," 69–71, and Henry T. Lewis, "Why Indians Burned: Specific versus General Reasons," 75–79, both in *Proceedings—Symposium and Workshop on Wilderness Fire* (USDA Forest Service, Intermountain Forest and Range Experiment Station, General Technical Report INT-GTR-182, 1985); Shepard Krech III, *The Ecological Indian: Myth and History* (New York: W. W. Norton, 1999), 101–122; Stephen J. Pyne, *World Fire: The Culture of Fire on Earth* (New York: Henry Holt, 1995), 303ff.; and Gerald W. Williams, "American Indian Use of Fire in Ecosystems: Thousands of Years of Managing Landscapes" (paper presented at the annual meeting of the American Ecological Society, Albuquerque, New Mexico, 12 August 1997).

2. Concerning the concept of "nature," see Kate Soper, *What Is Nature? Culture, Politics, and the Non-Human* (Cambridge, Mass.: Blackwell, 1995); William Cronon, ed., *Uncommon Ground: Rethinking the Human Place in Nature* (New York: W. W. Norton, 1996); and Michael E. Soulé and Gary Lease, eds., *Reinventing Nature? Responses to Postmodern Deconstruction* (Washington, D.C.: Island Press, 1995). Closely related to the question of the nature of "nature" is the concept of "wilderness." Two seminal works on this subject are Roderick Nash, *Wilderness and the American Mind* (New Haven: Yale University Press, 1982), and Max Oelschlaeger, *The Idea of Wilderness: From Prehistory to the Age of Ecology* (New Haven: Yale University Press, 1991).

3. Williams, in "American Indian Use of Fire in Ecosystems," 15, writes: "In some cases, the agencies will be faced with a public that has been so ingrained with the Smokey Bear ethic that taking fire back into ecosystems seems at odds with decades of anti-fire promotions. For others, the reintroduction of fire will appear to be a waste of a valuable resource (trees), while for some it will be a long overdue savior of the federal forests. In reality, it is none of the above."

4. Phil F. Brogan, *East of the Cascades* (Portland: Binfords and Mort, 1964), 234–245.

9. TURNING

The epigraph is from Walt Whitman, *Walt Whitman: Complete Poetry and Collected Prose*, comp. Justin Kaplan (New York: Literary Classics of the United States, 1982), 632.

1. Keith Clark and Lowell Tiller, *Terrible Trail: The Meek Cutoff, 1845* (Caldwell, Idaho: Caxton Printers, 1966), 83.

2. Aldo Leopold, *A Sand County Almanac* (Oxford: Oxford University Press, 1987), 228.

10. ELK CAMP

In the epigraph, Goudsblom (*Fire and Civilization*, 13–14) is summarizing the speculations of nineteenth-century anthropologist Karl von den Steinen.

1. Søren Kierkegaard, *Concluding Unscientific Postscript to "The Philosophical Fragments": An Existential Contribution by Johannes Climacus*, in *A Kierkegaard Anthology*, ed. Robert Bretall (New York: Modern Library, 1946), 193.

11. SETTLING

The epigraph is from Bachelard, *The Psychoanalysis of Fire*, 14.

1. Even science fiction recognizes and projects into the future this primordial valuing of fire tending. In the movie *Star Trek V: The Final Frontier*, we find twenty-third-century Kirk, McCoy, and Spock sitting around a campfire in Yosemite Valley, roasting marshmallows. In this

same film genre, in *Star Trek III: The Search for Spock,* we find what appears to be a fireplace in Kirk's San Francisco apartment. Surely, in such a future age in which even nuclear fission has become obsolete, hearths would hardly be needed for heating. And though it is not made clear whether the fire is real or holographic, its presence testifies to the (hypothetically) enduring value of fire itself.

2. Pyne, *Fire in America,* 69.

3. Matt. 3:12.

4. See Christian Hole, *Christmas and Its Customs* (New York: M. Barrows, 1958), 29–34.

12. BRACKETED BY FIRE

The epigraph is from Erazim Kohák, *The Embers and the Stars: A Philosophical Inquiry into the Moral Sense of Nature* (Chicago: University of Chicago Press, 1984), 25.

EPILOGUE

The epigraphs are from Heraclitus Fragment 37, in Kahn, *The Art and Thought of Heraclitus,* 45, and John Calvin, *Sermons on Job,* first sermon on chapter 37 (Edinburgh: Banner of Truth Trust, 1993), 675.

INDEX